FIRST LESSONS IN BEEKEEPING

DADANT PUBLICATIONS

American Bee Journal
American Honey Plants
Beekeeping Questions and Answers
Bees and the Law
Contemporary Queen Rearing
First Lessons in Beekeeping
Honey in the Comb
Instrumental Insemination of Honey Bee Queens
The Hive and the Honey Bee
The Honey Kitchen

First Lessons
in
BEEKEEPING

By C. P. DADANT

Revised and rewritten by

M.G. DADANT, J. C. DADANT,
DR. G. H. CALE JR., AND HOWARD VEATCH

DADANT & SONS · HAMILTON, ILLINOIS

COPYRIGHT, 1917, 1924,
BY C. P. DADANT

COPYRIGHT, 1938,
BY J. C. DADANT

COPYRIGHT©, 1976,
BY DADANT & SONS

Second printing, 1978
Third printing, 1979
Fourth printing, 1980
Fifth printing, 1981
Sixth printing, 1983
Seventh printing, 1985
Eighth printing, 1986
Ninth printing, 1987
Tenth printing, 1990
Eleventh printing, 1997

Library of Congress Catalog Card Number 75-38347

ISBN Number 0 915698-02-1

PRINTED IN THE U.S.A.
BY
GREAT RIVER PRINTING COMPANY
HAMILTON, ILLINOIS

BEEKEEPING IS PLEASURABLE AND PROFITABLE

KEEPING BEES is a fascinating and desirable pursuit. Steeped in age less time, it has survived since the early recordings of the Vedas in India, to be heralded as well in the Koran of Mohammed and the Holy Bible of Christianity. Equally it attracted the attention of Greek mythology, of Pliny and Aristotle and on down to the more modern investigations of Huber and Fabre.

The fascination of keeping bees lies in the fact that even the man of average talents can get a liberal education in the objectives and accomplishments of this cooperatively organized group in the beehive called the colony; while the specialist in practical beekeeping or the scientist never will succeed in learning all there is to be learned about this little sacrificial trudging honey gatherer.

Here is a coordinating group of three types of bees acting in unison for the perpetuation of the bee colony, wearing themselves out in the summer rush for the harvest of pollen and honey. Apparently dormant though continually active during the winter season, the colony maintains a compact winter cluster by food and energy in order that the mother queen and her retinue of workers may survive again to build for another season. Thus the race is perpetuated, for generation after generation, for century after century.

Within the confines of the hive the eggs are laid, and develop into drones, queens, or workers. The worker bees pursue in their turn: housekeeping duties, work in the field, or efforts in defense of the hive in case of attack.

Apparently without reasoning power and largely through instinct they communicate to each other when there is nectar in the field, in what direction it is and how far afield. Von Frisch, of Austria, has analyzed that peculiar tail-wagging, circular running activity of the worker bee within the hive as a definite communication of the fact that the time for action has come— nectar is available.

All of these and many more equally as interesting phenomena of the honey bee colony are available to the beginner. He need only secure a hive of bees and handle them properly to be highly repaid for his investment and time, in enjoyment of the unfolding of one of those mysteries of life which go to make up our universe.

Bee stings may be a deterrent to some, but to those familiar with the bee and her habits, they are only a minor consideration. Proper care in approaching the hive, gentle and proper opening and handling of the bee colony will minimize the possibility of getting stung.

Beekeeping offers a mild exercise or hard work according to its extent. It offers the reaction of the open air in the happy days from spring's opening to leaf fall of autumn, while it requires a minimum of attention and effort during the rainy days or the rigorous season of winter.

But there are other rewards of the products of the honey bee: the one, cross-pollination of fruits, legumes and vegetables, is intangible the others, honey and beeswax, directly and profitably apparent.

Honey and beeswax we all know about; the former was our earliest sweet, used universally until Alexander the Great returning from an Asiatic exploration, brought with him the first sugar cane. The latter, beeswax, equally as ancient, was used then as now as a medium of light; then, by necessity, now decoratively or in religious custom. Strangely, equally as prominent as the religious use of sacred candles, comes the use of beeswax for milady's complexion, for the face cream of today has as its base the nectar of the flowers transformed by the bees themselves into the beeswax of commerce and the substance of which their own combs—their own abodes are built.

Yet, sweet as is the honey of the hive, smooth and refreshing as is the beeswax of the cold cream, neither of these exemplifies the honey bee's greatest contribution. For as the bee hurries from flower to flower, sipping the nectar from the clover or the apple, the cucumber or the vetch, she simultaneously gathers on the hairs of her body and legs a bit of dustlike material, called pollen, from the flowers. True, much of this pollen is brought into the hive to be mixed with nectar to form the food of the baby bees. But not all of it. As the bee flies from one flower to the other, leaving a bit of pollen on each as she drifts, she aids in the perpetuation of those trees, shrubs, and plants, through cross pollination. (Fig. 1) .

Some of our plants are self-pollinating, others are wind pollinated, but some ninety or more of our vegetables, fruits, and legumes must have insect carried cross-pollination to assure not only a sufficient quantity of fruit or seed but also a satisfactory quality.

There probably was a time when our native beneficial insects could do the job of plant, shrub and tree perpetuation. However, our best authorities now recognize that the following combination of factors has left the honey bee our sole hope for future pollination: intense cultivation, large scale single crop acreages, and the immense destruction of beneficial as well as harmful insects through sprays and poisons. If we are to have adequate cross-pollination to secure good crops, ample and plump seed and fruit, and a perpetuation as well as improvement in our agricultural picture, these tireless little bees will be largely responsible. The recent awakening of our agriculturists to the need of soil conservation for the proper retention of soil content means more legumes, more seed for seeding those legumes, and

Fig. 1. A bee visits a blossom. Pollination will be a side-benefit of this visit. (Photo by Chas. S. Hofmann)

more bees to do the job of carrying those necessary pollen grains from one blossom to the other.

While it is undoubtedly true that the honey bee has been worth ten times as much in aid to pollination as for her products of honey and beeswax, it is equally true that our agriculture of the present and future requires the aid of our honey bees many times more than it did a hundred years ago.

BEEKEEPING IS UNIVERSAL

Wherever flowers bloom, bees may be kept. The success of their efforts, of course, depends upon the rigors and length of the season and on the amount of bloom available. While the vigorous extremes may preclude profitable bee culture in the Arctic or the desert, yet some bees are kept in

Alaska, and nomadic tribes carry their rude colonies with them in the Near East as they wander across the sands with the seasons.

Fifty years ago the Dakotas and the Western Canadian provinces imported honey. Bees apparently were thought unable to flourish there. Yet, today these very sections represent some of the best commercial honey producing regions of the American continent.

Nor has civilization and city congestion limited the possibility for keeping bees. For wherever there are flowers, honey bees may be kept.

BEEKEEPING MADE EASY

The beekeeper in antiquity knew little of the bee colony organization. His hives were rude log gums, crudely fashioned clay cylinders, or straw skeps. (These still persist in many countries of Europe, Asia and Africa and we too have our share of box or log hives and "gums" in certain American localities. See Fig. 2).

But we of present-day America may share the progress brought to beekeeping everywhere in the 1850-1870 period by the inventions of three men which have revolutionized the method and ease of keeping bees.

Rev. L. L. Langstroth discovered the significance of a bee space in the interior of the beehive and subsequently invented the movable comb hive, a

Fig. 2. In a combination of modern and antiquity, the log "bee gum" at the left has been outrigged with a shallow super. 2b shows box hives still in use in some parts of the world.

hive with frames made of wood, each of which contains a honey comb. The frame hangs by an extension of its top bar at both ends of the hive, leaving a 3/8 to 1/2 inch space at the ends and bottom of the hive which serves as a passageway for the bees and is called the bee space. This space allows room enough for the withdrawal of the frames from the hive for examination or replacement but not enough space between the frame and wall of the hive to induce the bees to build comb attachments. Langstroth's hive was made with the top removable so that any or all frames might be lifted independently from the hive body.

Following Langstroth's invention, Johannes Mehring in Germany conceived the idea of furnishing to the bees a part of the beeswax they needed in forming their combs. His idea was to imprint into a flat thin sheet of beeswax, the impression of the cells of a comb. Thus, all the bees need do is to draw out these cells of wax, adding beeswax of their own production to finish the proper depth of the comb. So, we now have straight combs instead of the conglomeration of immovable cross-combs as in the skep, box hive, or gum. These sheets fitted into the Langstroth frame would be drawn out by the honey bee colony, providing ease of handling and examining the colony and equal ease of producing and removing the surplus honey. (See Fig. 3).

Franz Hruschka in Italy, meanwhile, had discovered that combs of honey whirled centrifugally could be made to release their store of honey; and the third item for modern honey production was born with the invention of the honey extractor. (See Fig. 9, p. 30). No more need the combs be cut out and

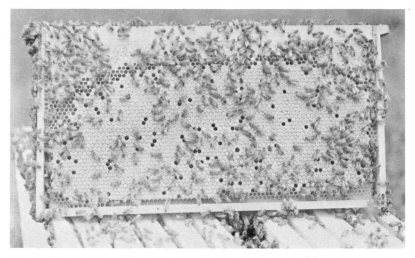

Fig. 3. The bees have "drawn" out the cells of this sheet of foundation and are using it for rearing young bees (brood comb). The cells at the top of the frame are being used for storage of honey and pollen.

squeezed to get the liquid honey. Merely slice off the cappings of full combs of honey, combs built on Mehring's foundation, in Langstroth's movable frame, place them in Hruschka's honey extractor with resultant clear liquid honey, free of all bees, larvae and pollen. And the beauty of it was, after the honey was extracted these combs could be returned to the colony to be refilled by the honey producing colony and used again, year after year.

These three major inventions, with improvements and variations developed through the years, form the basis upon which the ease of keeping bees and the handling of honey is practiced today.

The beginning beekeeper, whether his goal be that of mild exercise and pure pleasure, a hobby, sideline business or production on a larger scale, has at his disposal a wealth of information, both to get him started and to answer any questions that may occur along the way.

College courses in beekeeping are readily available, even correspondence courses, as well as lectures and short instruction classes in many vocational and educational schools. Slides and films are available from various outlets.

Available literature concerning bees ranges from basic books such as this one to more comprehensive texts such as The Hive and the Honey Bee,* used in most college courses. Trade magazines such as The American Bee Journal,* manned by staffs with many years of beekeeping experience ready to answer any personal correspondence, offer current articles of interest as well as advertisements of bee suppliers.

Bee suppliers, in addition to offering catalogs of complete beekeeping equipment, are another source of obtaining technical advice both for the beginner as well as the more advanced apiculturist.

Additionally, many states have extension apiculturists ready to help answer the questions of the beginner or the commercial beekeeper.

---

*Dadant Publications, Dadant & Sons, Inc., Hamilton, Illinois 62341.

CHAPTER II

WHEN, WHERE AND HOW
TO START WITH BEES

To ANYONE with an inclination toward nature study and the outdoors, beekeeping offers an occupation both fascinating and profitable. The keeping of bees has been and is being practiced by men and women in all walks of life, whether located in the city, a suburb, on a small farm or large acreage. The beginning may be made in a small way at a minimum of expense and with possibilities of early gain.

WHEN TO BEGIN

The beginning has already been made. Reading in this book indicates that the main ingredient for beekeeping success, curiosity about one of Nature's most fascinating insects, is already present. What remains now is to proceed in an orderly manner to gather the other essential ingredients for success—knowledge, equipment and practical experience.

Whether starting with a fully established colony or with new equipment and package bees, the best time is early in spring, usually about fruit bloom time. Thus, the package is able to build up to a full colony, sometimes before the major honeyflow is too far advanced.

Remember that the first season is the learning season. Learning how to manipulate the equipment, how to handle the bees, and how to diagnose and solve the problems of the bees should be the main preoccupation of the beekeeper while the bees will be busy fully establishing their colony. While one need not be disappointed if the colony only succeeds in fully establishing itself the first season, the honeyflows may be so bountiful as to give a harvest of fifty pounds, or even more, of honey with a good strong colony for winter quarters.

WHERE TO BEGIN

No particular location is necessary for keeping bees. Bees have a range of flight of over 12 square miles, which makes it possible to keep a limited number of colonies of bees in cities where they get their forage from flower beds, gardens, and the flowers of clovers and weeds of vacant lots within their flight range. Even the mountainside and the desert usually furnish sufficient nectar from their plants and trees to harbor a small number of colonies in a limited area, the number of colonies to be kept in one location or locality depending upon the amount of flora available and the nectar producing possibilities of the flowers.

Select the specific location for your colony or colonies so that the line of flight of the bees will not interfere with passers-by or with the neighbor's garden or yard activities. If the colony can be placed under trees, the foliage of which is not too thick, so that the bees may make their exit directly up through the branches, the chances of difficulty with the neighbors will be lessened.

Wherever the location, raise the beehive from the ground a few inches on blocks to avoid dampness and if there is a good windbreak on the north and west in the form of a fence or shrubbery, take advantage of it. Usually hives are faced with entrances away from prevailing winds. A south or east exposure is good.

HOW TO BEGIN

How will the beginner get his start with bees? Three ways are open to him: He may purchase a full colony of bees in good condition from a neighboring established beekeeper. He may secure box hives or bee tree logs and do his own transferring into the modern movable-frame hive. The beginner may also purchase his beehives and equipment from a bee supply firm, prepare them for the bees, then import packages of bees from the South for his beginning.

If the start is made by purchasing a full colony of bees, preferably in spring, from a neighboring beekeeper, some handling equipment will have to be acquired. The advantages of starting by this means are: (1) the bees will already be established, and (2) through the purchase you will have acquired the friendship of a fellow beekeeper who may be a good source of information. The main disadvantage here is that you may also run the risk of failure of an old queen, or of bee diseases.

Some beginners may be located where box hives or log gums are available for a nominal sum, but again the vigor of the stock may well be under par, disease may be present, and the necessity of transfer of the bees and combs to new movable frame hives presents considerable difficulty for the beginning beekeeper.

Probably the safest and most efficient way to get started would be through the following schedule:

1. Read. Become familiar with bees, what they do and how they live and organize their colony. Become acquainted with various types of hives, supers, and other bee equipment (Chapter III); as well as with the inner workings of the colony itself, its various members, their structure and their duties as outlined in Chapter IV.

2. Acquire a beekeeping hobby kit, preferably during the winter months. This will provide an opportunity to assemble the equipment, paint the outside of the hive and learn to manipulate the various parts of the hive, and become familiar with the equipment. Experiment with the gloves. hive tool, veil and the smoker. This rehearsal will be invaluable when the beginner opens his first hive. (See Chapter III for illustrations.)

A good hobby kit is shown immediately below:

Fig. 1. This hobby kit contains everything that the beginner will need with the exception of the bees.

3. Purchase a package of bees with a young laying queen. This method has the advantage that the beekeeper knows exactly what he is getting for his money. The bees are well bred. The package consists of a young queen, young worker bees, and few drones. They are packed for shipment in a light screen cage and come to your door by parcel post, with a minimum shipping cost. As they are reared in the South during early spring, they may be delivered to points in the far North in ample time for the earliest honeyflows. This gives them a full season to establish themselves and perhaps to yield a good crop during the first year.

Package bees can be obtained from nearly all bee supply houses which carry, in addition to beginner supplies, protective clothing, additional hives and supering kits and materials, complete lines of extracting and packaging-aids as well as all preventive medications and of course, their valuable advice.

The best way to begin is with two or three colonies rather than one. With a single colony, though the chances of failure are rare, the beginner may be discouraged in such instances whereas the second colony might make a bountiful harvest, to offset the rare loss.

There are bees of several races, the most prominent of which are the Italian, the Carniolan, and the Caucasian. The Italian race has been most generally accepted as the standard in America and should be your beginning choice. Later, other races may be tried if desired.

Line breeding and hybrid crosses are rapidly being developed in bee breeding programs. In such programs as the Dadant Starline and Midnite Hybrid bee breeding program, good attributes as gentleness, high honey gathering qualities, color, disease resistance etc. are combined into strains of bees of superior quality.

Later in the season, when the bees start making more honey than is needed for their own requirements, storage room will be necessary. This is provided by frame supers with either thin foundation for cut comb honey or heavier weight foundation for extracting. Section comb honey may also be produced at this point by comb honey supers with section boxes and thin surplus foundation. However, the beginner would be wise to stay with cut comb honey or extracted honey production. Section comb honey production is a highly specialized art that requires a considerable knowledge of bee behavior and apiary skill.

Fig. 2   a   b   c
Three of the most common European Races of Bees. a. Dark Bee (Apis mellif-era mellifera L.) b. Italian Bee (Apis mellifera ligustica) c. Carniolan Bee (Apis mellifera carnica).

CHAPTER III

BEEKEEPING EQUIPMENT

A NY DISCUSSION of beekeeping equipment would be premature without a word or two about the bees who will live in, manipulate, and in their turn, be manipulated by this equipment. Bees may be successfully kept in nearly every area and condition that the world has to offer with the exception of extremes of heat and cold and minimum standards of vegetation, but they will never be wholly domesticated nor fully trained. A dog will generally do his master's bidding because the dog has accepted his master as a literal master and depends upon him directly for food and shelter and gives in return, varying degrees of affection and companionship. The bee neither gives nor takes such liberties with its independence.

Our best research would indicate that bees are motivated by factors such as stages of development, genetic composition, and internal and external stimuli to the body. The processes of intelligence and thinking and decision-making as we know them, are not significant factors if even present at all.

To simplify the matter, if the bees find that the food, supplies and equipment provided by man are to their convenience, they will probably make use of them. If not, they will rely upon inborn impulses and make the best of the situation in which they find themselves, whether it makes any sense to us or not.

The history of beekeeping is littered with equipment that was ultimately rejected by the bees and the beekeeper, however inventive or insistent has always been forced to wait for and rely upon the final judgment of the bees. The items briefly described in this chapter have received this seal of approval and have passed into the industry as standard items of equipment.

The Hive

A good hive gives the beekeeper complete control of the combs which are the heart of his business. In order to reduce crowding and minimize swarming, the hive must be large enough to accommodate a prolific queen and all of her brood. Additionally, it is in this hive that the bees must live, work, raise this brood, and store their two surplus foods—pollen and honey.

Successful beekeeping means easy manipulation of the frames of brood and honey to provide a surplus of honey beyond that which the bees need to live on and rear their replacements. It is this surplus of honey which the beekeeper removes and markets for his product.

With the movable-frame hive all combs can be taken out, examined, and replaced or exchanged with those of other hives at will without drastically disturbing the work of the bees. The combs having a surplus of honey can be emptied by the extractor without injury and returned to the bees to

Telescoping Cover.

Inner Cover.

Extracting or
Bulk Comb Supers
(May be added as
needed).

Queen Excluder.

Brood Chambers
(One or two may
be used.)

Bottom Board.

Hive Stand.

Fig. 1. Relative positions of the component parts of the hive.

be refilled, thus saving the labor of the bees in making new combs. The queen can be found, examined, and when necessary, replaced; and new colonies can be made by dividing. If a colony is weak it can be strengthened by giving it a frame or two of brood from some other hive or it can be fed by supplying it with combs of honey from wealthier colonies. In short, the movable frame enables the beekeeper to control the condition of his bees and their increase.

COMPONENTS OF THE HIVE

Hive Stand As the bee is a flying insect, the hive stand with its angled landing strip, is a convenience for bees returning to the hive loaded with honey or pollen. Usually constructed of durable cypress wood, a primary function of the hive stand is keeping the hive off the damp ground and keeping the cluster and combs drier in winter. (See Fig. 1, p. 18).

Bottom Board As the hive bodies themselves will have neither top nor bottom, a bottom board is necessary and is contained in the beginning kit. This bottom board serves as the floor of the beehive and is supplied with various means of reducing or enlarging the entrance to the hive. It is made with a shallow side reducing the entrance in winter and a deep side for summer providing better ventilation. (See Fig. 1, p. 18)

Hive Body This is a rectangular box, dovetailed at the corners for a weatherproof fit, and equipped inside for suspending the movable frames. The best size to use is the standard 10-frame Langstroth style hive body. Chief among the many advantages of using the standard Langstroth hive body is that all the hive parts will be interchangeable, an important fact to keep in mind for future expansion. This hive body will be variously called: hive body, brood chamber, or full-depth super, depending upon its use in the hive.

The first hive body, resting on the bottom board, will be the brood chamber where the queen lays the eggs and the baby bees are raised. Two brood chambers are recommended for each colony, the second brood chamber with its frames of foundation being given to the bees when the frames of the first are nearly completely drawn. (See Fig. 1, p. 18). For the beginner, looking for a definite guideline, supply the second brood chamber when the first contains seven or eight frames of fully drawn comb. Later, experience will perhaps dictate other guidelines.

When the second brood chamber is perhaps half drawn with comb, it is time to begin the process of supering by adding the shallow supers where the honey will be stored in comb cells.

Queen Excluder Looking rather like a grill removed from an oven, the queen excluder, if used, is usually placed between the brood chamber and the supers. Its construction indicates both its purpose and its name. The size of the grillwork prevents the queen from passing into the supers where the honey is stored, while allowing the worker bees easy passage in either direction. The queen is restricted to laying eggs only in the brood chambers and the upper supers will be free of both brood and pollen. (See Fig. 1, p. 18).

Supers   Rectangular boxes the same outside dimensions as brood chambers, the supers are constructed in varying depths for different reasons such as ease of handling and varying types of production.

1. **Full-depth super** The full-depth super, using the 9-1/8' depth frame, may be used for production of extracted honey. The advantage here is that many beekeepers, especially those operating commercially, are sufficiently concerned with standardization and interchangeability, that both brood and supering areas, where the honey is stored, will consist of nothing but hive body depth. A disadvantage is that it will weigh approximately 85-90 pounds when full of combs of honey, presenting handling difficulties for some beekeepers.

2. **6-5/8" Dadant depth super** This super, also known as a shallow super is used mainly for the production of extracted honey in the 6-1/4" frame. The finished weight will be approximately 35 pounds.

3. **5-11/16" Shallow super** This super uses the 5-3/8" frame for the production of bulk comb honey as well as extracted honey.

4. **4-13/16" depth Comb honey super** This super uses the 4-1/2" frame size for bulk comb honey or holds 28 of the 4-1/4 x 4-1/4 x 1-7/8" sections for section comb honey production. Section comb honey production is a specialized skill and requires additional equipment for its production such as section holders, separators, super springs etc. usually supplied in kit form from bee supply houses.

Combinations of the above supers may be used on one colony, depending upon the type of honey production desired. All are readily accepted by the bees if properly used and each size and style of super will call for varying means of handling and methods of management.

Inner cover A rectangular cover, usually made of masonite surrounded by a frame, is supplied with the beginner kit and fits between the top hive body and the telescoping cover or roof of the beehive. The inner cover has a rectangular hole with rounded ends cut into the center which may be equipped with a bee escape.

Bee escape The bee escape is a practical item of beekeeping equipment even though it does its work rather slowly. It is a small metal device in two parts that slide together like a match box. On the inside the bee escape resembles a fish trap in function. Bees enter the escape through a round hole in the top, travel through flexible prongs extending from the sides, and exit through either end. The prongs allow the bee to go through the escape but do not permit a return.

When used in removing supers of honey it is inserted in the center hold of the inner cover which is placed under the super or supers to be removed. The bees are unable to return to the supers after traveling down through the bee escape to reach the brood chamber or hive entrance. By lifting the supers in the evening and placing the bee escape under them, the supers will be free of bees the next morning with the exception of a very few bees who remained with the honey. Advantages of the bee escape include a sure way of removing bees from supers without direct handling of the bees. Disadvantages include time involved, the fact that two trips to the hive must be made instead of one, and the fact that unless the supers fit tightly above the bee escape, robber bees may find and pilfer the unprotected honey before the beekeeper becomes aware of his loss.

Telescoping cover This is the roof of the beehive and is supplied in beginner kits. It is usually metal covering a wooden frame for added protection against the weather and its sides telescope well down over the inner cover and the top super or body for a rainproof fit and extra stabilization in high winds. (See Fig. 1, p. 18).

The above, constituting the component parts of the hive, are all available from beekeeping suppliers. The beekeeper should consider carefully his needs and projected type of honey production before investing. Other factors to be kept in mind are: 1. the interchangeability of the components and, 2. protecting all exposed surfaces of the hive with a good coat of paint to preserve the wood and aid in control within the hive.

FRAMES
The Hoffman or self-spacing frame is in general use. It is particularly

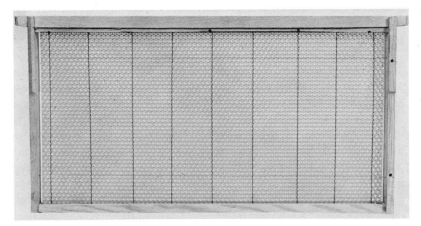

Fig. 2. The Hoffman frame with a sheet of Dadant crimp-wired foundation installed.

desirable for beginners because they cannot make the mistake of putting too many or too few frames in each hive. Frames containing the combs of the bees are spaced from 1-3/8 to 1-1/2 inches from center to center within the brood chamber. Closer or wider spacing will result either in narrow and imperfect combs or in the building of two combs on the same support, making undesirable irregularities. Loose hanging frames are preferred by some beekeepers as they are more easily handled in the apiary and more easily uncapped. The all-important requirement in movable frames is that the comb be built straight in them. All frames are made with devices for holding the bee comb foundation in their top bars by a wedge which is nailed fast or a groove in which the foundation is fastened. (See Fig. 2).

BEE COMB FOUNDATION

This is the beeswax foundation based on Joseph Mehring's discovery, mentioned briefly in Chapter I. These imprints or foundations of the cells are formed of beeswax in varying sizes and thicknesses for varying types of honey production. Space restrictions do not permit a discussion of all types of foundation and a beekeeping supply house catalog should be consulted for specifics in types, dimensions, and uses of foundation.

As a general rule, specific use will dictate the weight of comb foundation to be used. Weight is usually determined by how many sheets of foundation of a given size may be manufactured from one pound of beeswax. Foundation will be heavier and thus, sturdier, for use as brood comb and extracted honey produced in the full-depth super and lightest in weight and the most fragile for use in comb honey sections.

To further increase the strength and to permit a sheet of foundation to be used for either repeated brood rearing or extraction of honey as well as any moving or transporting as might be necessary, some manufacturers reinforce the wax sheet in a variety of ways.

Reinforced comb foundation for brood or extracting purposes first became practical in 1921 when Dadant & Sons perfected a method of wiring which consisted of vertical crimped wires woven into the foundation by machinery. (Fig. 2). Crimped wires prove better than straight ones for the shoulders of the crimp radiate reinforcement between the wires and thus, throughout the sheet of wax, and prevent the beeswax from slipping downward when soft from heat. Nine or ten vertical crimped wires are ample to prevent sagging and generally, two to four horizontal wires are used in addition to the vertical as a support to prevent the foundation from swinging in the frame. This double wiring results in a rigid straight comb and permits rapid handling and long distance hauling with little or no damage to the combs even though newly built and heavy with honey.

In 1963 Dadant & Sons carried the reinforced foundation concept a step further by introducing a plastic base comb foundation that has since become extremely popular. A thin film of plastic material between layers of beeswax is embossed with the honeycomb cell and is readily accepted by the bees. Called Duracomb,® or Duragilt® with a metal strip along either short end, this foundation is convenient to assemble and has proven to have a long and durable lifespan under constant usage. (See Figure 3).

Fig. 3. Dadant Duragilt® foundation. The two holes in the bottom of the foundation are called "communication holes" and will be used by the bees as they travel through the combs.

Foundation used for production of bulk comb honey will be thinner than that used for brood comb or extracted honey and the sheets of foundation used for the production of comb honey will be as thin and delicate as possible to avoid toughness due to a heavy midrib. Foundation wax for comb honey will be made of the best grade of light colored beeswax, such as is obtainable from rendering cappings and new combs, for the appearance of the product is also of importance here. By comparison, the wax used for milling the foundation used for brood-rearing combs can be of darker beeswax, usually obtained from rendering old combs and frame scrapings.

Whatever its source, foundation beeswax must be refined until free of all impurities for two reasons: one, the bees show a decided preference for pure beeswax and may reject all other types, and perhaps even more important, secondly, the major uses for beeswax are other than in comb foundation manufacture such as candles, cosmetics, pharmaceutical and industrial products where absolute purity is a must.

The advantages derived from using comb foundation are threefold. In the first place, beeswax costs the bees a probable average of eight pounds of honey to produce each pound of comb. The less wax the bees have to produce for themselves, the greater the savings both in time and honey for the beekeeper.

The second advantage, which is equally important, is that it gives the bees a guide on which to work. Before the advent of bee comb foundation, guides of different kinds were devised to compel the bees to build straight in the center of the frames or sections. In spite of all efforts, the combs were often crooked or wavy and irregular. With comb foundation as a guide combs are built straight in every frame and section. This advantage alone would pay for the cost of the foundation.

The third advantage is almost as great as the other two. In natural conditions the bees build about 10 per cent of drone comb. This is necessary in a state of nature when colonies are far apart and the queens in their mating flights need to meet drones readily. But as only a few drones are actually able to service one queen, the number of drones of one first-class colony is quite sufficient for fifty or more colonies in one apiary. Hence, it is advisable to do away with as much drone comb as possible. By using bee comb foundation embossed with worker cell bases, large areas of drone comb are dispensed with and replaced by worker comb. There will always be plenty of drones reared in corners of the frames or in cells that become enlarged by accident. It must be remembered that the drones do no work in the colony and the annual savings by the prevention of rearing a horde of useless consumers through the use of worker comb foundation is sufficient to pay for the initial cost of this foundation.

Once the particular type of honey production desired is established, the appropriate foundation is secured in its frame and given to the bees. The bees "draw" or work this basic imprint of a cell into the fully-drawn cell by adding bits of wax of their own production. Collectively, these cells will be known as comb, whether they will be used for brood rearing or any type of honey production.

This cell will be hexagonal in shape, have three four-sided rhomboids as a base, and represents the ultimate in economical construction. It will have surprising strength and capacity, will take up the least amount of space of any structure, and at the same time, consumes the least amount of labor and wax in its construction than any other possible shape. In a further economy measure, the base will be used on both sides; that is, each facet or rhomboid of the base of one cell, will provide a third of the base of a cell "drawn" on the opposite side of the foundation. A final feature of this magnificent structure is that the walls of the finished cell will slope upward as it hangs in its frame in the colony, so that the stored honey will not spill. See Figures 4 and 5.

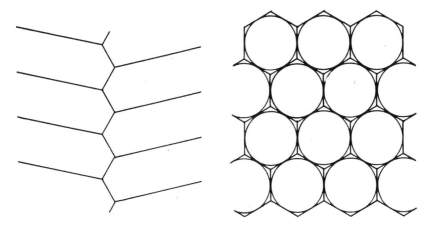

*Figs. 4-5. 4. Diagram showing slope of cells from front to middle of comb.
5. Diagram showing the economy of the hexagonal shape for making honeycomb cells*

The Beekeeper's Equipment

The previous sections of this chapter dealt with the equipment necessary for the bees. Certain items are also necessary for the beginning beekeeper. While there are beekeepers who handle their bees without the ben-
efit of veil, smoker, gloves, etc. and who may wear anything from suits to
shorts, they are probably not beginners. This book will strongly recommend
protective clothing and handling equipment. An individual style of management and handling of bees should be based only on considerable experience.

The bee is a stinging insect and although it loses its life in the process
of stinging, it will not hesitate to do so if sufficiently aroused. One sting
produces an unpleasant irritation for the beekeeper and a dead bee. The following items of equipment, illustrated in Figure 6, are designed to prevent
both.

Light-colored, preferably white, clothes are best for beekeeping. Black
or woolen clothing should not be worn in the apiary. Wool is a fuzzy tex-
ture of animal origin while cotton is of vegetable origin resembling the
stems of plants among which the bees seek their substance. A man dressed
in a light-colored thin cotton suit or pair of coveralls, even with the sleeves
rolled to the elbows, is safer from stings than if he were dressed in the thick-
est of woolen garments. Likewise, bees seem to find felt hats particularly
irritating and a lightweight helmet is a good investment.

A good pair of gloves is necessary for the beginner. A variety of elbow-
length canvas, plastic-coated canvas, or leather gloves are obtainable from
beekeeping supply houses and should be worn until familiarity brings loss

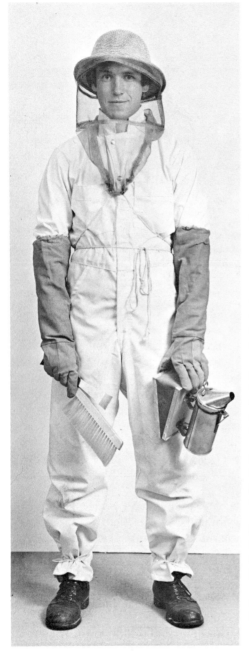

Fig. 6. *This beginning beekeeper will be able to manipulate and observe his hives at leisure, knowing that he is protected against stings and well-equipped to handle any situation.*

His white coveralls, tied at the ankles, protect him and yet allow maximum freedom of movement.

The woven fiber helmet is cool and comfortable and supports the wire and cloth mesh veil away from the face.

The full, elbow-length sleeves on the gloves offer maximum protection. Although the experienced beekeeper may discard the gloves, they are helpful to the beginner until he becomes accustomed to an occasional sting.

The smoker will be used to produce a calming effect on the bees. When smoke is applied, the bees gorge themselves with honey and are then easily handled.

The hive tool is an all-purpose tool for nail-pulling, prying, scraping burr-comb and propolis and indispensable in handling the colony.

The soft bristles of the bee brush will be used to gently brush the bees from clothing, hive parts, or frames of brood or honey.

Each of the items shown are discussed more fully in the text of this chapter.

of fear. After fear has been overcome, the gloves may be unnecessary. They are cumbersome at best and the advanced apiarist will want complete freedom of his hands at all times.

A bee veil such as the one shown in Fig. 6 should be worn by the beekeeper to protect the face and neck from stings. There are several types of veils available but the veil which stands out from the face and is stiff enough to keep the wind and tree branches from brushing it against the face, offers the best protection. A good veil worn in conjunction with a helmet is a lightweight, cool, and worthwhile investment in protection.

BEE SMOKER

The bee smoker, shown in Fig. 7 and supplied with the beginner kit, is one of the most important tools the beekeeper uses. A smoker may be simply described as a small hand-held stove containing a low smouldering glow of a fire and is equipped with a bellows for injecting air into the firebox which produces a short puff of smoke from the nozzle. Rolled corrugated paper, frayed clean burlap or partly rotted wood may be used for fuel. A comparatively heavy but cool smoke is desired.

Fig. 7. Smokers in two sizes. Note grate and firepot, with bellows attached at side.

The smoker works to subdue bees because it frightens them and they rush to gorge themselves with honey preparatory to flight. Bees, like men, are much more docile with a full stomach than an empty one.

A puff or two of full rich smoke at the entrance, followed by similar puffs over the frames of the hive as the top or supers are removed, is a great aid in keeping bees gentle and facilitates handling the colony without getting stung. Use a minimum of smoke, just enough to subdue the colony;

with an occasional puff or two over the tops of the frames as you proceed with the examination. Too much smoke or too hot a blast is apt to make the colony "run" even to the point of all congregating on the front of the hive and in the air around it.

Hive Tool The hive tool is another inexpensive but necessary item of equipment. It is an all-purpose tool used for pulling nails, taking off the cover, inner cover and supering and other utility purposes It is required for separating the different stories of the hive and for prying the frames apart since they are always more or less glued together by propolis. It is also valuable for scraping away burr-combs that are often built by the bees between the frames or between the supers when the hive is crowded.

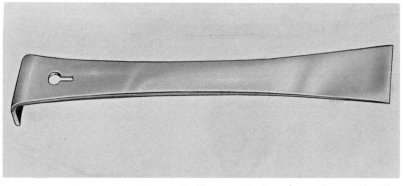

Fig. 8. 10 inch (25.4 cm) hive tool. Sharpened edges for prying, nail puller with good leverage.

Bee brush The bee brush has long fiber or soft plastic bristles and is ideal for removing bees from clothing, hive parts, and frames of brood or honey. Vegetable fibers irritate bees less than animal material and the soft fibers prevent injury to the bees.

As the operation expands, the beekeeper may wish to purchase a bee blower, basically a motor driven impeller in a housing generating a stream of air applied through a flexible hose and nozzle.

Miscellaneous Equipment

All sorts of feeders are manufactured. Some are simple while others are more complicated. All should be used with care and like any other item of beekeeping equipment, experience and experimentation on the part of the beekeeper will determine which type of feeder is best for any given operation.

Entrance Feeders The entrance feeder is supplied in the beginner kits and consists of a standard screw-on lid with perforations that fits any size mason glass jar. The jar, filled with syrup, is inverted over a holder, one end of which fits into the hive entrance. An advantage is that its use does not require opening the hive and the level of its contents may be checked without disturbing the bees. A disadvantage is that any type of outdoor feeding tends to excite the bees and may lead to robbing, where the weak colony being fed may be overpowered by stronger colonies even though they may have excess stores of their own.

Division board feeders This is a plastic container resembling a frame that can be filled with syrup and hung in the hive replacing a frame. An advantage is that bees can remove feed even in cold weather. A disadvantage is that it takes up the space that a frame of honey might occupy.

Spur embedder The spur embedder closely resembles its name and is used to embed the horizontal wires into foundation to add to its strength. A rotating wheel with teeth as in a riding spur is set into a handle and passes over wire, embedding it into foundation. The embedder may be used heated (preferably) or cold and two tools (one heating while the other is in use) will speed up the operation.

Honey Handling Equipment

Honey handling equipment needed depends upon the type of production desired. Here the beginner has a decision to make. He may decide to produce honey in the little pound boxes called sections and consume or dispose of it in section or comb honey form. In such cases, he will order comb honey supers with comb honey sections and foundation to fit.

Or he may wish to produce for his own use in shallow frames (5-3/8 or 6-1/4 inches deep), comb honey which he can cut out of the frames for home consumption, to give to friends, or to sell wrapped in cellophane. In such case, he uses shallow supers with frames and thin foundation. Thus, the combs may be cut out when filled, new foundation inserted, and the frames returned to their supers for the balance of the crop.

The beginner may ultimately wish to produce liquid or extracted honey. In such case, the same shallow supers may be equipped with a reinforced foundation, which is heavier; or added hive bodies with full depth brood frames may act as storage space for surplus honey. This extracted honey production, however, will require additional removal equipment such as an uncapping knife and honey extractor.

Perhaps the first season, in any case, the beginner would be better off as his colony builds up, to order his surplus honey storage space for either comb honey or for bulk honey in shallow frames. The latter has the added advantage that such storage equipment may be used for extracted honey production if a change to that type of production is made later.

For the producer of extracted honey, an uncapping knife for cutting away the capping surface of the comb is necessary. These knives are made in various styles, either plain or steam or electrically heated. Likewise honey extractors for whirling the honey from the comb after it has been uncapped may be had in a wide range of sizes.

Fig. 9. 2-Frame, basket type, hand-operated extractor. Centrifugal force throws exposed honey from the cells onto the sidewall, then down into a collecting tank.

Wax Extractors

Manufacturers of bee comb foundation require much of the available beeswax in the country; every bit of wax and old combs should, therefore, be saved. A wax extractor will soon pay for itself, for with it all old combs and cappings may be saved, rendered into beeswax and eventually restored to the bees as comb foundation.

It is not always absolutely necessary to have a wax press or wax kettles, for an ordinary wash boiler may be used, though with more waste than a wax press manufactured for the specific job. Soft water should be used in melting wax. Iron utensils are objectionable because the iron rust colors the wax and soils its appearance permanently. Tin or tinned receptacles are indispensable. For very small amounts, a double-boiler is usable.

Break up the combs, soak them well in water, then heat them to the boiling point, taking care not to overboil the wax, as it would spoil it and perhaps cause it to run over. Make a sort of basket or pouch wire cloth and sink it into the surface of the boiling mixture with the combs submerged beneath it. From this you may dip the wax as it comes to the surface and pour it into flaring crocks or tin pans. The few impurities that you may dip up will settle to the bottom and the wax may again be melted to finish cleaning it

As many people do not like to trouble themselves with rendering old combs into wax, the bee comb foundation manufacturers have taken it up on a large scale. In many cases the apiarist can have his old combs rendered and made into comb foundation for less than he could do it, especially as waste is avoided.

The beekeeper gathers, from time to time during the season, a large amount of bits of beeswax from old combs or from scrapings of hives frames and sections. The solar wax extractor is inadequate for rendering large lots of combs, although it is quite handy to dispose of small amounts of wax. Its cost is not great and the broken combs thrown into it are thereby protected against the moth and may be gathered at the end of the season.

Observation Hive

No man can keep bees successfully unless he becomes well acquainted with their habits. To a certain extent he may acquire knowledge of bees by reading; but he cannot thoroughly understand bees, unless he is acquainted with them by actual contact. Therefore, one should try to study the habits of bees and for this, an observation hive is desirable.

A good observation hive is composed on only one comb in a movable frame in a hive the sides of which are made of glass. This hive has either doors to cover the glass or a black cloth cover. The doors are better excluders of light but the opening and closing of them often jars the bees slightly and disturbs them. The hive may be placed in a window with an entrance at the front so the bees may forage. It is usually stocked in the spring by taking a good comb of brood from one of the best colonies with plenty of bees to keep the brood warm. The first thing the bees will do is to rear a queen.

Fig. 10. Observation hive. To start, add comb with adhering bees, queen and provide feed. This model has small upper level to produce comb honey.

It is possible to observe the different changes through which the brood passes, the emergence of the queen, the bringing in of pollen and honey, etc. It is an endless source of interest and instruction. At the end of the season, as it is difficult to winter bees in so small a hive, it is advisable to unite with some populous colony. In the spring the observation hive may be restocked with bees and brood.

CHAPTER IV

THE HONEY BEE COLONY

T HE HONEY bee colony is made up of three individual types of bee—the the worker, queen and the drone. These three carry on the functions of the colony and in its usual working condition, a colony of bees contains a fertile queen, many thousands of workers, according to the season of the year, and in the busy season from several hundred to a few thousand drones.

Fig.1   a. Drone   b. Queen   C. Worker
The distinctive sizes and shapes of the three types of bees are useful in telling them apart in conditions such as as the shadowy interior of the hive.

THE QUEEN

The queen is the only perfect female in the colony and is the true mother of it. Her only duty is to lay the eggs for the propagation of the species. She is a little larger than the worker but not so large as the drone. Her body is longer than that of the worker but her wings are proportionately shorter and her abdomen tapers to a point. She has a sting but it is curved and she uses it only to fight or destroy rival queens.

When she is about five or six days old the queen takes flight to mate with drones outside upon the wing. After being fertilized she remains so for life, although she often lives three or four years. A few days after mating she commences to lay and she is capable, if prolific, of laying as many as three thousand eggs a day. These are regularly deposited by her in the cells within the brood chamber of the hive. The queen usually lays from January to November, but very early in the spring she lays sparingly. When fruit and flowers bloom and the bees are getting honey and pollen she lays most rapidly.

Fig. 2. The queen (center) surrounded by her "court" of attendant workers bees. While the queen looks for cells in which to lay her eggs, workers encircle her as she moves over the combs, touching her with their "feelers", grooming and feeding her.

The ovaries of the queen, occupying a large portion of her abdomen, are two pear-shaped bodies composed of many tubes. The eggs originate in the upper end of the tubes, pass finally into the vagina and out through the oviduct.

The spermatheca is a globular sac which contains the male semen. All eggs pass by the spermatheca duct but not all eggs are fertilized. Each egg which in passing receives one or more of the seminal filaments, produces a worker or queen, while an unimpregnated egg produces only a drone.

If for some reason the queen is unable to mate within the first three weeks of her life she loses the desire for mating but is nevertheless able to lay eggs. These produce only drones. Also, if the drone's organs were sterile or their supply exhausted, or if she has been rendered infertile by being chilled, a queen may lay eggs which develop only as drones. In either case such a queen is, of course, worthless and should be replaced.

The ability of a queen to lay unfertilized eggs which develop into adult bees belongs only to a few species of insects and is called parthenogenesis, discovered in queen bees by Dzierzon.

It is necessary to have a queen that is prolific and should she become barren from any cause, or become lost, or even decrease in her fertility during the breeding season, or die from old age or from accident, the worker bees immediately prepare to rear another to take her place. This they do by building queen cells out and around worker cells which contain queen larva.

By feeding the embryo queen with royal jelly, the egg that would have produced a worker, becomes a queen. Royal jelly is probably a misnomer though it is used by most authors. Jelly in plentiful supply is given to the queen larva during the entire time of its growth.

The bees also rear queens when preparing to swarm. The first emerging queen destroys the others and the bees usually help her to do it, unless they wish to swarm again. With the exception of her mating flights, the queen leaves the hive only when accompanying a swarm.

THE DRONES

The drones are shorter, thicker and bulkier than the queen; their wings reach the entire length of the abdomen. They are much larger and clumsier than the workers and like the queen and workers are covered with short, fine hair. Their buzzing when on the wing is louder and differs from that of the workers.

They are the males and have no sting, neither have they any means of gathering honey, secreting wax or of doing any work that is necessary even to their own support or to the common good of the colony. Their only use is to serve the queen on her bridal trip. The drone loses his life in the act of copulation, dying instantly. Not more than one in a thousand is ever privileged to perform that duty, but as the queen's life is very valuable and the dangers surrounding her flight are numerous, it is necessary to have a sufficient number of drones in order that her absence from the hive may not be protracted. That is why hundreds and often thousands of drones are reared by the bees in each colony in the spring during the breeding and swarming season. In domestication, when dozens and sometimes hundreds of colonies are kept in an apiary, the choice colonies alone should be made, by giving them drone comb, to rear drones in large numbers for reproduction.

After the swarming season is over, or should the honey season prove unfavorable and the crop short, the drones are mercilessly destroyed by the workers. (Fig. 3). Should a colony lose its queen, the drones will be retained longer, since without the drone the young queen would remain infertile and the colony would soon become extinct.

When comparing the head of the drone with those of the queen and the worker one readily notices the compound eyes, those crescent-shaped projections on each side of the head. They are much larger in the drone than in either of the others and this is ascribed by scientists to the necessity for finding the queen in the air on the wing. The facets composing those eyes number some twenty-five thousand in the head of the drone, so that they can see in all directions. The three small points, present as well in queen and worker, in a triangle at the top of the head are small eyes, or ocelli, which are probably used to see in the dark within the hive and at short range.

Fig. 3.   This drone will soon by excluded from the hive (Photo by E. R. Jaycox)

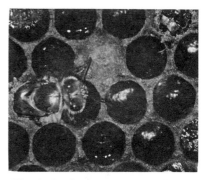

Fig. 4. In the center, a capped worker cell.   At left and upper right, worker bees are emerging from cells. (Photo by J. B. Free)

Fig. 5.   This worker bee may be feeding upcapped brood or cleaning the cell for its next occupant.   (Photo by Richard F. Trump)

Fig. 6.   These honey bees are collecting water from a pool for use in the hive.

Fig. 7.   A worker comes in for a landing while on a nectar-gathering flight. (Photo by Ben M. Knutson)

Fig. 8. A pollen-gatherer approaches a flower. Note the pollen grains lodged in the body hairs. (Photo by J. B. Free)

THE WORKERS

The worker bees are underdeveloped females and they do all the work that is done in the hive. They secrete the wax, build the comb, ventilate the hive, gather pollen for the young and honey for all, feed and rear the brood, and fight all the battles necessary to defend the colony. Of the three kinds of bees these are the smallest but they constitute the great mass of the population. They possess the whole ruling power of the colony and regulate its economy.

Anatomically, the body of the bee is divided into three segments- head, thorax and abdomen. The details of the head of a worker bee are very interesting. We have already mentioned, when speaking of the drone, the compound eyes, which are larger, containing a greater number of facets in the male than in either the queen or the worker. The bees have short, thick, smooth mandibles working sidewise, instead of up and down as in higher animals. These mandibles, unlike those of wasps and hornets, have no teeth and yet enable the bees to perform their necessary hive duties and to mold the wax to build their combs. They are therefore incapable of cutting the smooth skin of sound fruits of any kind. The tongue of the honey bee is made of several parts: ligula, palpi, and maxillae. The central part, of ligula, is grooved like a trough; when at rest it is folded below the mentum, or chin. In the head and thorax are three pairs of salivary glands. The largest pair of glands is supposed to be used in the production of food for the larvae, as will be seen further. The antennae, or feelers, are the two long "horns" which protrude from the head of the bee. These exist in all insects. The popular name of feelers is proper, for it is with these antennae that the bee examines everything with which it comes in contact. They appear to serve the purposes of smell, touch and hearing. As there are usually tens of thousands of bees in a colony and they very readily recognize their own members, it must be with the antennae that this recognition is achieved.

The organs of breathing are in the thorax and in the abdomen between the rings or segments of the third section of the body.

The honey sac, or first stomach, is located in the abdomen or third segment of the body of the bee. From this stomach the bee may, at will, digest a part of the honey by forcing it to the second stomach for the nourishment of its body or it may be discharged back through the mouth and stored in the honeycomb cells for future use.

The honey bee has four wings and six legs all fastened to the thorax or second segment of the body. The wings in pairs fold upon each other to enable the bees to enter within the cells where the brood is reared and where the honey is stored. In flight the two sections of these wings are braced together by very fine hooks, which cause the wings to present a greater surface in contact with the air. It is not necessary to go into the details of the different segments of the legs of the honey bee, but it is well to say that each leg is supplied at its extremity with claws, which permit the bees to hang to each other in the cluster. Near the claws there is a small rubber-like pocket, which secretes a sticky substance. This enables the bee like the fly and other insects to fasten itself and to walk with ease upon any smooth surface such as a pane of glass or a ceiling. The anterior legs are provided with a notch and a thumblike spine which is used by the insect to cleanse the antennae. The motion made for this purpose is often noticed in house flies as well as in bees. The third pair of legs of the worker bee have a hollow portion, called the pollen basket, which enables the bee to carry home the pollen used to make the food for the young. This pollen is popularly called bee-bread. Neither the queen nor the drones are supplied with these pollen baskets. They would have no use for them since they never work in the field.

The ovaries, or egg pouches, which are very large in the queen, are almost absent in the workers, which are therefore incomplete females and unfit for mating, although they may occasionally be able to lay a few eggs which mature as drones. On the other hand, the sting, which in queens is curved and which they use only as an ovipositer and to fight other queens, is straight in the worker and is accompanied by a much more highly developed poison sac. The sting is barbed and is used by the workers for self-defense and for the protection of their colony. Unlike the queen, which may wield her sting again and again, the worker uses her sting only once; for in so doing she injuries herself and soon after loses her life.

The worker may live as long as a few months or more in the winter, when she is not flying about; in summer her life is very short averaging less than forty days. She literally wears herself out. For that reason a queenless colony, in which the number of bees is no longer replenished by bees emerging daily, soon dies. A colony which has failed to raise a queen after swarming or whose queen has been lost in her wedding flight will be entirely depopulated by fall. Those reared in the fall, having little outdoor work to perform, will live till spring. None of them die of old age, but the majority work themselves to death and many die from accidents.

Put the workers, the queen and drones together, give them a hive allow them sufficient time to build their combs and you have a colony. The number of bees in this colony may vary all the way from a few thousand to sixty thousand or even more. Bees may be described as social insects, that is, they band together and have a division of work among themselves.

DEVELOPMENT OF THE BROOD

The brood is the eggs and young of the bee. It is to be found in the combs in the brood nest, or body, of the hive. The student will readily learn to recognize the three different kinds of brood. The queen cell, when empty, looks like an acorn cup and when sealed like a peanut. Sealed worker brood differs from sealed honey in its uniformity, and in the leathery appearance of its cappings. The drone brood differs from the worker brood in its roundish cappings, which look like bullets.

Fig. 9. a. The peanut-shaped queen cell built outward from a worker cell like a room added to a house. b. Worker cells, some capped, others ready for sealing. c. Bullet-shaped drone cells.

There are three stages in the development of the bee, whether queen, drone, or worker, before it becomes a perfect insect. These stages are egg, larva, and pupa. The queen deposits the egg in the bottom of the cell, in three days it hatches into a small white worm called a larva, which, being fed by the bees, increases rapidly in size. When the cell is nearly filled by the growing larva, it is closed up by the bees. The larva then enters the chrysalis or pupa state.

Royal jelly is fed to the queen larva during the entire time of its growth and to the worker and drone larvae during the first three or four days, after which they are fed a coarser food, which is a mixture of pollen and honey.

The worker develops from the egg in twenty-one days; the drone reaches the adult stage in twenty-four days; the queen matures in fifteen days. In ordinary circumstances the worker bee does not leave the hive until about eight days after maturity. During that time she is not idle. She serves as a nurse to the larvae and at about the eighth day she takes her first flight. A number of young workers usually take flight together in the warmth of the afternoon. They fly about the entrance in happy and humming crowds, making peaceable circles to learn the location of their home that they may return to it without error. It is of some importance for the beginner to learn to recognize the first flight of young bees, as the bustle of their pleasant sally

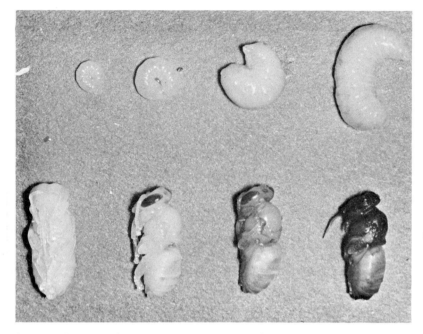

Fig. 10. The successive changes in development from the egg, through the larval and pupal changes. (Photo by J. B. Free)

somewhat resembles the actions of robber bees with which it must not be confused. The following table gives the duration of development of the brood from the egg to the perfect insect:

		Queen	Worker	Drone
In the egg ....days		3	3	3
Growth of larva ....days		5	6	6-1/2
Spinning of cocoon ....days		1	2	1-1/2
Period of rest ....days		2	2	3
Change to pupa ....days		1	1	1
Change to winged insect ....days		3	7	9
Average duration of changes .....days		15	21	24

These figures represent the time required in summer weather, in cool or cold temperatures the time is somewhat lengthened.

WAX PRODUCTION AND COMB BUILDING

The two products of commercial value made by the bees are beeswax and honey. The production of wax, which the bees use for building their combs, is one of the most remarkable phenomena in the organization of the honey bee. The segments or rings of the abdomen of the bee, six in number, overlap each other. At the underside of four of these rings are pairs of five-sided, transparent surfaces on which small scales of wax are formed by a peculiar process of digestion. Each worker bee is able to produce wax; queens and drones are not.

The production of wax is involuntary whenever the bee is compelled to remain a long time with its stomach full of honey. In amount it is imperceptibly small in ordinary circumstances but increases rapidly as soon as conditions demand it. As long as there are plenty of empty cells in the hive to receive the crop, the bees are not compelled to retain honey constantly in their stomachs and there is only enough wax produced to repair or to elongate the cells and to seal them. But as soon as the want of room compels many of the bees to remain filled with honey for twenty-four hours or more, a sufficient amount of wax scales is produced to build combs to store the surplus honey.

When a swarm is about to leave its old home and seek another, each bee fills itself with honey. Having entered their new home, the gorged bees suspend themselves in festoons from the top of the hive. They remain motionless for about twenty-four hours. During this time the honey has been digested and converted into beeswax.

As colonies usually swarm only during a good honeyflow, many of the bees composing the swarm often have scales of wax already produced under their abdominal rings. This explains the great speed and apparent cheapness of production of comb in such circumstances.

The cells in which the worker bees are reared measure about five to the inch or a trifle over twenty-seven to the square inch. The cells in which drones are reared measure four to the inch or about eighteen to the square inch. The total for both sides of the comb is, of course, double that number.

A number of cells called intermediate cells are built when the bees are changing from worker to drone comb. These intermediate cells, or cells of accommodation, are of irregular shape and size, varying between the other two according to requirements. Besides the cells already enumerated, there are large cells which hang downward and are shaped like an acorn or a peanut. These are found placed here and there, especially at the edges of the combs. They are the queen cells. In them the queens are reared for swarming or to replace the old queen when she becomes infertile. The worker and drone cells are used not only for brood rearing but also for storing honey. Pollen is almost invariably stored in worker cells.

Fig. 11. Worker bees "drawing" foundation into comb, using bits of beeswax which they have produced.

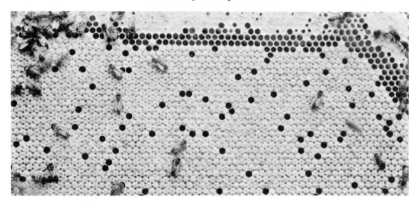

Fig. 12. The finished cells have been put to good use. The capped cells in the center contain "brood". The cells at the top contain honey and pollen.

The thickness of worker combs is about an inch, with space for passage of bees of about seven-sixteenths of an inch down to five-sixteenths. As these distances may be slightly increased without troubling the bees, many beekeepers place the combs in the hives one and a half inches apart from center to center. This spacing increases the ease of manipulation.

It is estimated that from seven to fifteen pounds of honey are required to be consumed by the bees to produce a pound of comb. The quantity undoubtedly varies greatly according to the conditions in which the bees find themselves when the comb is built. The greatest amount of comb is secured during a strong honeyflow in a summer temperature. Excessive heat is objectionable in this connection only when it softens the wax too much and causes a breakdown.

At first when the combs are built they are white, but with age and use of brood rearing they become dark and opaque. The thin cocoons lining the cells help to make them dark; they are, however, just as valuable for breeding purposes for a long time, or until the size is materially diminished. They are valuable, because of their toughness, when the honey extractor is used. During a harvest of honey and pollen of deep yellow or amber shade the comb promptly assumes that color, although it was white when first secreted by the bees.

HONEY

Honey is a vegetable sweet produced mainly from the nectar of certain flowers. Nectar is not made by the bees but only gathered, although during the transfer from the blossoms to the hive in the stomach of the bee it undergoes a slight change by the action of enzymes in the bee's body. Thus it is customary to call it nectar before it is gathered and honey when it has been placed in the cells. It contains more or less water according to the season in which it is produced, the atmospheric conditions, the amount of moisture in the soil and the plant from which it is harvested. It is said that the fuchsia produces exceedingly thick nectar. Nectar gathered from the heather in dry sandy plains is often so thick that the honey is extracted from the comb with difficulty. On the other hand, white clover honey is often so thin when first gathered that it drops like water from the combs when they are handled by the apiarist. It sometimes contains from 75 to 90 per cent of water. But the bees soon ripen it in the warmth of their hive with the aid of ventilation which they give it by the fanning of their wings. During a heavy harvest, rows of bees may often be seen facing the hive, extending from the edge of the alighting board into the hive and all through its combs, their wings moving with such rapidity that they are invisible. This serves not only to help to evaporate the honey but to give pure air to the inside and to keep down the temperature of the hive.

The quantity of honey harvested by a colony of bees in a single day may vary from a few ounces to twenty pounds or more. But a portion of this gain may be promptly evaporated as the honey thickens. Twenty-five per cent of the weight of fresh nectar usually disappears during the first twenty-four hours. When the honey is sufficiently ripened, the bees seal or cap their cells with a thin cover of beeswax.

In addition to the nectar of the blossoms, the bees sometimes gather and bring to their hives a sweet substance called honeydew, which is mainly produced by aphids or plant lice. Honeydew is of very low quality and not properly acceptable as honey. But since the beekeeper cannot avoid the gathering of it by his bees during some seasons, its addition to the total honey crop tends to lower the quality.

POLLEN

Pollen is the fertilizing dust of flowers. The pollen which the bee gathers and helps to spread upon the pistil to fertilize the seed is needed in the economy of the hive. It serves to make the nutritive food given to the worker and drone larvae in the latter part of that stage in their lives. During the winter, however, when the bees are inactive and not rearing brood the need of pollen is not felt. At that time honey only is suitable for food and that which contains the least quantity of pollen is most healthful for the bees, for it leaves very little residue in their intestines during the long confinement. Honeydew, fruit juices, and dark honey loaded with pollen grains are bad winter food.

Let it be remembered that bees are needed for the pollination of most fruits, many of the vegetables as well as many of the clovers, alfalfa, vetch and other legumes, as mentioned in the first chapter of this book. Not only do they cause the pollen to be spread upon the blossoms which they visit, but in their flights they carry it from one blossom to another and thus bring about cross-pollination and greater production of fruit and seeds.

PROPOLIS

Propolis is a resinous substance gathered by the bees from the buds or limbs of some trees. It is brittle in cold weather but so sticky in the summer that the bees apply it immediately to the purpose for which they secure it-to stop the cracks in the hive, to reduce too large an entrance, and often to strengthen the combs at their junction with the walls. They use it also to cover obstacles which they are unable to remove, such as partly rotten wood in hollow trees which they may inhabit, or to cover and in some manner to embalm the bodies of large insects or the bones of mice which have found their way into the hive and have been stung to death. Bees carry propolis to the hive in the pollen baskets. During dull, dry summers large quantities of propolis are sometimes gathered and the entire inner walls of the hive are thickly coated with it.

WATER

At all breeding seasons, when fresh, watery nectar is not available, water is necessary to the bees in preparing the food given to the larvae. In early spring bees are often noticed around pumps and watering troughs, unless some stream close at hand gives them a constant supply of water. Water is not needed for the adult insect, but it is well to see that it may be had within a short distance of the apiary during cool spring days when the lives of the workers may be endangered by long trips. A bucket with floating chips is the handiest water trough, as the bees are in no danger of drowning.

CHAPTER V

PACKAGE BEES—
FROM ARRIVAL TO HONEYFLOW

In CHAPTER II, package bees, complete with their young, laying hybrid queen, were recommended as the easiest, most trouble-free way of getting started with beekeeping.

Let's assume that the beginning beekeeper has read some background literature on bees and beekeeping, has chosen and contacted a beekeeper supply manufacturer, and has acquired a beginning beekeeping kit. An order for live bees and their queen will have been placed well in advance of the first season and the beekeeping kit has been assembled and the exposed surfaces of wood given a good coating of paint. The basic equipment is ready—it is time to go into business.

Package bees are as the name implies, a package containing live worker and drone bees and a young queen. They are grown in and shipped from

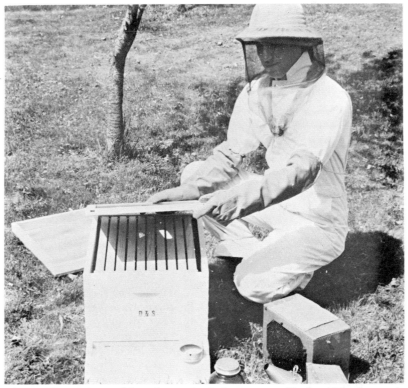

Fig. 1. The package of bees has arrived, the entrance block and feeder are in place, and the feeder jar is filled with sugar syrup. It's time to open the hive and put the bees in.

the southern tier of states early in the spring and should arrive around the time of early fruit bloom (apricot and pear and ending with apple bloom some four to five weeks later). They will arrive in either wood and screen-wire or corrugated cardboard and screenwire cages by parcel post. Usually each package contains two pounds of bees, although occasionally three pounds per package are shipped. Inside the package a young queen is confined to a separate small queen cage suspended by a wire, and a small can of sugar syrup is included to feed the bees enroute.

Many beekeepers feel it is advisable to give the bees an additional feeding as soon as they arrive. The feed is a sugar-syrup solution made by mixing one part of granulated sugar to one part of warm water. It is fed by dipping a clean paint brush into the solution and gently brushing the screenwire sides of the cage from which the bees will feed. When the bees quit picking up the syrup readily, they have had enough and the package should be placed in its normal upright position until all is ready to actually install them in the hive.

There is no particular rush about installing the package of bees after it has arrived, provided there is an adequate supply of food. However, It is best to place the package in a cool, dry- and preferably dark-room, such as a basement, where they are fed sugar syrup until hived.

The next step is to locate the equipment (bottom, hive body with its frames of foundation installed, inner cover, outer cover, feeder and entrance reducer) where the colony of bees is to be kept. Ideally, the best time to install a package of bees is in the late afternoon or early evening. This is to prevent the bees from making any flights the first day, before they have had a chance to become organized within the hive and to mark the location of the hive in relationship to the surrounding area where they will forage.

Fig. 2. The best time to install a package is in the late afternoon or early evening. Make sure all equipment is ready for use and in place before opening the package as installing the package should be a smooth, continuous operation.

Fig. 3. Wet the bees thoroughly with warm water. This can be done with a hand sprayer or you may dip a large paint brush in a bucket of warm water and sprinkle the water on the bees. This wetting prevents the bees from flying.

Fig. 4. Using the end of the hive tool, the next step is to pry up and remove the cover from the top of the package. Be sure to keep the cover near you because there is need for it when the feeder is removed.

Fig. 5. After removing the cover, take the sharp end of your hive tool to lift up and remove the small feed can that was shipped with the package of bees The remaining syrup in this can may be added to that which you have prepared for the bees.

Fig. 6. When the feed can is removed, place the cover over the opening in the top of the package. This prevents the bees from getting out of the package while you are disposing of the can.

Fig. 7. You will notice that one end of the queen cage has some white candy in it. The cork covering the hole which leads to this candy should be removed and a small hole, about the size of a match, punched through the candy.

Fig. 8. Now suspend the queen cage in the hive. The proper place for the cage is three of four frames in from the side of the hive. Be sure that the end containing the candy is toward the bottom of the hive.

Fig. 9. Take the package cage containing the bees and bounce it on the ground. This will jar the bees to the bottom of the package cage. Pour about half of the bees in the package directly over the frames where the is suspended.

Fig. 10. The next step is to jar the bees into the bottom of the package again and pour all of the remaining bees possible over the frames. The bees on the top of the frames will find the queen almost immediately and begin to eat the candy and release her.

Fig. 11. There will still be some bees left in the package. Place the package with the remaining bees in front of the colony and make sure that the hive contains all of its ten frames with foundation.

Fig. 12. Next, place the inner cover on the hive. This should be done very gently so that the bees are not crushed. A little smoke on the top of the bees will make them rather quickly run down between the frames and make placement of the inner cover easier.

Fig. 15. The next step is to place your telescoping cover on top of the inner cover again being careful not to crush bees. The bees are now safely housed in their new home.

Fig. 14. Your colony has a lot of work to do. To help them secrete wax and build combs, be sure to use your entrance feeder and to keep it full of sugar syrup for at least the first six weeks.

15. The entrance feeder jar is carefully placed in the entrance feeder that was supplied with your colony. Note the package cage laying in front with some few bees still clinging inside. They will rather quickly crawl out and join their sisters within the newly established colony.

Fig. 16. The last step in hiving your package is to lightly stuff the small entrance with a little green grass. This confines the bees to the hive for a short time and allows them to become accustomed to their new home before they take flight.

Fig. 17. The job is done, your colony is now safely housed and fed in its new home. Success or failure will now depend on the care that you give the developing colony between the time of package hiving and the beginning of the first honeyflow.

In addition to the above equipment, the beginning beekeeper should also protect himself. Installing a package of bees calls for concentration and careful movements and the distraction caused by a sting could be a problem. A bee veil and gloves are necessary. Shirt sleeves should be long and tucked inside the long protective sleeve of the glove and trouser legs should be tied at the ankles. A hive tool and a lighted smoker should be ready for use.

This is a good opportunity for the beginner to examine the queen. Some shippers include five or six attendant worker bees in each queen cage to feed and care for her enroute while others send only the queen, relying on the workers to feed her through the screenwire sides of the queen cage. Note her size and appearance in relation to the workers. It may be necessary to locate her in the hive later on. If she has been marked with a small dot of paint on the top of her thorax, locating her will be much easier.

One end of the queen cage will have some white food (candy) in it and that end of the cage will also have a hole in it that will be covered with various types of material, paper, metal and corks, for example. It is necessary to remove this covering unless it is paper, so that the bees can eat through the candy and release the queen. If the candy has become very hard, it may be necessary to make a hole through the candy with a small nail to make it easier for the bees to release the queen. The idea behind confining the queen upon first installing the bees is to allow them to become organized and set up housekeeping inside the hive without becoming disorganized and swarming out of the hive. The bees will not swarm without the queen, or if they do, they will return to the hive shortly, as to swarm without the queen would lead to their loss.

The foregoing instructions have been keyed to general situations. There will be exceptions with every situation. No two beekeeping operations are exactly alike and that is part of the interesting challenge mentioned earlier in this book. It is the beekeeper's job to diagnose and solve the problems of the bees so that they can get on with their work.

As an example of exceptions, some shippers do not include candy in the queen cage and she will have to be released upon hiving the package or she will starve to death, as she is ill-equipped to forage for herself or even feed and care for herself should the food be readily available. Generally, the bees will accept the queen readily, whether they have to eat through the candy to release her or she has been released upon hiving the package. The acceptance process relies to a great extent upon smell. The queen has an odor that attracts the bees and their natural tendency will be to release her, feed her and care for her and immediately set to building cells in which she will begin to lay eggs which will hatch into more worker bees. Another natural tendency of the bees is to protect the queen from harm as the success of their colony depends directly upon her health and well-being.

The wise beekeeper will recognize that he too has now become an enemy, or at best, an intruder, and will realize that the bees will not be especially grateful to him for having supplied them with such a nice hive to live in, and will take precautions, both for himself and for the bees.

Let the bees alone for the first few days. They have thousands of years of nature's teachings and instincts to rely upon. Disturbing the hive will excite the bees and during these first few critical days, such a disturbance could result in the death of the queen, either through "balling" - clustering so tightly around her that she is wounded or smothered, or in the opposite-complete rejection of her.

HOW TO MANAGE THE PACKAGE
UNTIL THE HONEYFLOW BEGINS

The day after the bees are placed in the hive, examine the entrance to make sure that the bees have been able to work their way through the grass placed there. If the bees have not yet been able to get out of the hive, loosen the grass slightly, but do not pull it out. **Be sure that you do not remove the cover of the hive or disturb in any way.** It will take the bees some time to settle down in their new home. They will have to either release the queen from the cage or become organized around her if she was released immediately, but in any case, they will have to accept her. They will also have to build comb on the foundation in the frames and all of these operations will take time. During this first week the colony must have ample feed so make sure that the feeder jar contains syrup at all times.

One week from the day the package was installed it is safe to open the colony and examine it. By this time the queen should have been released and accepted and will probably be laying eggs in the newly built comb and the bees should have made some progress toward building comb on the foundation. In building this comb the worker bees take the wax scales that are secreted on the undersurface of their abdomen and shape and form them on the foundation until the wax bits conform to the pattern of the cells. This process is called "drawing" out the foundation and is an expensive process for the bees in that excess food and energy is used. Make sure that there is a constant supply of food available. Do not take food away from the bees simply because flowers are blooming nearby. Let the bees judge whether or not they need syrup. There are many factors which influence nectar secretion and an abundant bloom does not necessarily mean an abundance of nectar. When there is enough bloom so that they may secure nectar from flowers they will no longer need the sugar syrup and will no longer take the syrup from the feeder jar.

In making this first one week examination, the beekeeper will be establishing a routine procedure. Wear the veil and the gloves and secure both

Fig. 18. *A puff of smoke at the hive entrance will help calm the bee before opening the hive.*

Fig. 19. *The outer cover has been removed, the hole in the inner cover smoked, and the inner cover pried up with the hive tool.*

Fig. 20. *This hive has a honey super which is removed and the top of the brood frames is smoked.*

Fig. 21. *The top of the frames is smoked and a frame is gently pried loose.*

Fig. 22. *This frame of brood is removed, examined, and will be set aside so other frames can be removed.*

Fig. 23. *With the first frame set aside, the other frames can be removed for examination.*

wrists and ankles of clothing to prevent flying bees from entering clothing which will likely result in a sting. Approach the hive from the side. Approaching a hive from the front excites the bees even more and one of their prime jobs is to defend their hive against intruders. Carry a hive tool and a lighted smoker. Do not make any excessive noise nor any sudden movements. The object is to create as little disturbance as possible.

Remove the outer cover and gently smoke the hole in the inner cover. Smoking the bees subdues them in that at the first sign of smoke, the bees go to the unsealed cells and fill their abdomens with honey. They are then much easier to handle as they tend to be much quieter on the combs. Perhaps they are merely more docile with a stomach full of honey; perhaps they find it slightly more difficult to curve the abdomen into a stinging position with a full stomach. No one knows the exact reasoning or reactions of the bees but they are easier to handle and much quieter if the smoker is used with care. The smoke should be a cool smoke and the minimum amount necessary to calm the bees should be used.

Pry up the inner cover with the hive tool. It will often be glued to the top of the frames with propolis. Smoke the top of the frames. Find the suspended queen cage and if the queen has been released, remove it from the hive. Then lift out one of the center frames which faced the suspended queen cage. If you see eggs or larvae you can be sure that the queen is present and laying. Do not bother to look for the queen. Replace the frame, the inner cover, and the outer cover as quickly and as quietly as possible-but with easy motions.

Three weeks from the day you hived the package you may again make an examination of the colony. Be sure to refill the feed jar during every examination. During this three-week examination check again to make sure that the queen is present and laying well. By this time there should be considerable sealed brood in the combs. Do not attempt to change the position of any of the frames in the hive at this time. Three weeks after the package has been hived is the period of lowest population in the colony. Some of the bees which were shipped in the package will have died and none have yet emerged from the cells to take their place. Toward the end of the **fourth week** young worker bees will start to emerge from the cells, and during the fifth week they will emerge in even greater numbers. From the fifth week on the number of worker bees will increase fairly rapidly.

During the **fifth week** you should conduct another examination of the colony. Check again for the queen and make sure that the brood nest is expanding and that the bees are drawing their foundation well. It will help if you will move sheets of foundation that have not been drawn, up next to the brood nest. A word of caution here-always put frames with empty comb or foundation **up to** and **not into** the brood nest. If the frames of the brood nest are separated there is danger of some of the larvae being chilled if the weather is cold.

By the end of the **sixth week** the fruit bloom is over and clover bloom is close at hand. It would be well to talk with neighbor beekeepers and find out from their experience just when you may expect the main honeyflow to start. Usually by the end of the sixth week you may stop feeding syrup. There should be plenty of bloom present by that time to supply natural sources of food for the bees. If in doubt about the continued feeding, let the bees decide. They will stop taking syrup when they have other sources.

The package of bees is now a full-grown colony, humming with activity and ready for the honey flow. With a bit of luck and some judicious care on the part of the beekeeper, the bees will do the major part of the work. They will gather nectar from the flowers, transform it into honey, and store it away for the coming winter. If it is a good season, they will produce far more honey than they will need to overwinter. The beekeeping industry is, thus, a partnership in sharecropping. For their work, the bees will have provided themselves for the coming winter. For his work and expenditure for supplies and equipment, the beekeeper can remove and enjoy his share of the largesse.

Fig. 24. The fruits of your own labor. A moment of relaxation, a good cup of coffee, hot rolls and honey from your own hives.

CHAPTER VI

SOURCE OF BEE PASTURE

T HE EXTENT to which profitable beekeeping operations can be followed in any locality will depend upon the plants within flying range, their time of blooming, quality of honey which they yield and the acreage within reach. Honey for family use can be produced almost anywhere flowers bloom, even in large cities, but commercial honey production is only profitable where large acreages of heavy yielding nectar sources are present.

In most localities surplus honey comes from only a few sources and usually within a short period of time. Plants which support the bees before and after the main honeyflow are of equal importance since their presence or absence will determine whether the bees are ready for the flow when it comes. Some areas which provide heavy main honeyflows are poor beekeeping territories because there is no supporting flow in advance to enable the bees to build up strong colonies. Since the household duties of cleaning and ventilating the hive, building combs, feeding young bees, and guarding the entrance require a considerable number of bees, there can be but a small field force in a weak colony. It is only the strong colonies that have enough field bees to gather a worth-while harvest.

Where there are flowers within reach, the field bees will be bringing in supplies whenever the weather permits them to fly. From the time the first skunk cabbage or witch hazel opens in late winter until frost cuts down the last asters in autumn the bees will be seeking nectar and pollen to feed the young brood.

Although we are unable to see any accumulation of stores in the hive as the result of these early or late visits, they are of very great importance to the success of the beekeeper.

Where there is a large field force the bees are often able to store a substantial surplus from very short honeyflows such as come from black locust or the even shorter blooming honey locust. Weak colonies can only make use of such flows to increase the population of the hive.

When colonies are weak they often use a heavy honeyflow, such as comes from the tulip tree, for building up the colony instead of storing surplus. Thus, they build up on the honeyflow instead of for the honeyflow as has been noted by some of our leading authorities.

In regions where the honeyflow comes in midsummer there is likely to be a favorable opportunity to build strong colonies before the flow. In the Southeast where the flow comes early as it does from tulip tree, commonly called tulip popular, it requires expert beekeeping to get the bees ready in time for the flow.

IMPORTANT POLLEN PLANTS

Pollen which provides the protein for the growing larvae is consumed in large quantity and without an abundant supply there can never be profitable commercial honey production. In a country as large as this, with so great a variety of plants, it is difficult to give a proper outline of the subject which will be useful everywhere. In the accompanying chart an attempt is made to present the important bee pasture for the season. Many exceptions will appear in even these smaller areas. Plants which are given as minor sources of nectar for the region as a whole may be the main source of surplus in some sections. Other plants, which in some sections serve as early support, may yield a surplus in another locality. Only general statements are possible for such large areas.

Over a very large part of the United States the maples are among the first to yield either nectar or pollen. In limited regions plants like skunk cabbage may come earlier. The willows are very important for early spring support in all the northern states but in many southern localities they come into bloom much later, in some places as late as June. The native elm trees bloom very early in spring and yield a great abundance of pollen. In many places they are the principal source of early pollen.

As spring advances a wider variety of plants come into bloom and nectar and pollen both are more readily available. Too often the weather is rainy or cold and the bees are confined to the hives when the nectar harvest is on. At this season the fruit trees and the dandelions offer an abundance of forage and when weather is favorable, surplus honey is sometimes stored by strong colonies.

SOURCES OF SURPLUS—SWEET CLOVER

There are only a few plants which yield surplus honey in large quantity that are widely grown. There is a much larger number of plants which provide good crops in limited areas, but honey from these sources are seldom found in the markets of the large cities.

Sweet clover is probably the source of more marketable honey than any other crop. It is at its best in the Midwest, as it requires a rich limestone soil and low humidity. The honey is very light in color and mild in flavor. The common varieties of sweet clover are biennials. They make a vigorous growth the first year and come into bloom the second season after which they die. There are annual varieties, the best known of which is the white flowered "Hubam."

The yellow flowered sweet clover blooms a few days earlier than the white and in neighborhoods where both are grown extensively there is likely to be a honeyflow of extended duration which insures large yields in favorable seasons. Sweet clover is grown from Texas to the far north in western Canada. It is drought resistant and often yields good crops of honey in seasons too dry for satisfactory development of staple grains.

Fig. 1. Sweet clover is a midseason source of nectar. The honey is light in color and of a very mild flavor. It may be listed as the nation's main source of surplus honey.

ALFALFA

Alfalfa is also the source of large quantities of white honey from the mountain regions of the West. In the lower altitudes of the Southwest, the honey from alfalfa is darker in color and more highly flavored. It yields nectar freely on rich soils with abundant moisture at the roots and a dry atmosphere. In some seasons alfalfa is reported as yielding in the East but it is not dependable in humid climates. In seasons where a wet spring is followed by dry, warm weather the nectar often flows freely for a time until the soil becomes dry.

Most markets pay a premium for white honey of mild flavor such as comes from sweet clover and alfalfa. The large acreage of these legumes offers bee pasture of such extent as to make available this high quality honey in car lots in the principal markets.

There are few other plants sufficiently common to support the large apiaries now commonly operated by commercial honey producers. Outfits operating 1000 or more colonies are now common in the states where these plants are widely grown.

Fig. 2. BUCKWHEAT
    Most of the buckwheat honey is produced in the region of the Great Lakes in Ontario, New York Michigan, and Pennsylvania. Buckwheat does best on sandy or light soils and seldom yields on clay or heavy soil. Buckwheat yields a very dark honey of pronounced flavor which is popular with many. It is so different from the mild-flavored light honey from the clovers that it seldom appeals to the same people.

Fig. 3. SAGE
    Large amounts of surplus honey are secured from sage in California. There are three species of importance: the white sage, the black sage, and the purple sage. Sage is practically unknown as a commercial source outside of California where cultivation is rapidly reducing its range in more settled areas. Sage honey is white, of heavy body, and of fine flavor. It is thought by some to be the finest honey obtainable.

Fig. 4. GOLDENROD
    Goldenrods are widely distributed with more than 80 species known to North America. Honey in surplus quantity is reported from Louisiana. Goldenrods are of most value in some of the northern states in eastern Canada and in the New England states where they are the source of a deep golden honey of heavy body and decided flavor, granulating rather quickly but usually regarded as safe for winter stores for the bees.

Fig. 5. MESQUITE

The most important source of honey in Texas and the Southwest is the mesquite. Mesquite was once the most common tree from Texas through New Mexico to Arizona and eastern California. Over large areas it was the main source of honey for commercial apiaries. The honey is of light amber color and good quality. In areas where the land remains uncleared much fine honey still comes from this plant.

Fig. 6. FIREWEED

Fireweed has yielded big crops for a short time following the removal of the forests in New York, Michigan, Wisconsin, Minnesota and later in British Columbia, Washington, and Oregon. When the trees are cut, fireweed appears and covers the ground with a vast sea of pink flowers of great attraction to bees. The blooming period is long and fireweed offers wonderful pasture. The honey is light in color and of fine quality.

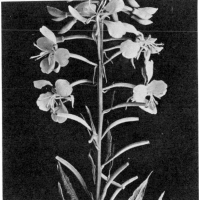

Fig. 7. HEARTSEASE

The heartsease and the related "smartweeds" occur over most of the United States and Canada. They are very common in late summer in stubble after grain is harvested and in cultivated fields where they continue to grow after cultivation has ceased. There is a great variation in quality as well as in color of heartsease honey, but it is usually of pronounced flavor and of light amber color.

WHITE DUTCH CLOVER

Fifty years ago the white clover was the principal source of white honey. It grows over a wide expanse of country from the Missouri River to the Atlantic Coast and southward. Much honey still goes to market from white clover but it no longer holds first place since the sweet clovers and alfalfa have been planted over such a wide area. Honey from white clover is slightly darker in color but of mild and pleasing flavor. Alsike clover yields well in neighborhoods where dairying is important or where it is grown for seed. The quality of the honey is similar to white clover. Neither yield as dependably as in former years before soil fertility had been reduced.

Fig. 8. ALSIKE CLOVER

Alsike was brought to this country about 1839 and was met with widespread acceptance. The beekeepers were largely responsible for its spread although the farmers quickly saw the value of this clover and adopted it. Alsike is grown principally in the dairy regions of Minnesota and Wisconsin, and to a less extent throughout the northeastern states and eastern Canada. Alsike clover honey is similar to that from white Dutch clover.

MINOR SOURCES OF SURPLUS

It would require more space than is available here to describe the long list of plants from which surplus honey comes in limited areas. In the Southwest the mesquite yields a light amber honey of good flavor. In the region from south Texas to southeastern California mesquite is important to the beekeeper in desert areas.

In the Southeast the sourwood tree yields a honey which is so popular in the region from which it comes that little goes to outside markets. Heavy yields are reported from the Carolinas and western Tennessee. The tree is found from West Virginia to Georgia and west to Arkansas but the big crops are seldom reported outside a limited area.

In Florida heavy yields come from tupelo along the Appalachicola River, from black mangrove along the east coast and from the palmettos in the interior.

California boasts of her orange honey which is famous far beyond her borders. Sage also yields heavy crops. Surplus is reported from a variety of sources including blue curls, eucalyptus, manzanita, lima beans, incense cedar and several others.

Fig. 9. ORANGE

Although the bees gather honey from a large variety of fruit blossoms, only the orange is a major source of surplus honey. While some honey comes from it in Florida, Texas, and Arizona, the bulk of orange honey is harvested in California where it is one of the principal sources of surplus. Orange honey is light in color and of marked but pleasing flavor, and is usually much in demand.

Washington and Oregon report surplus from cabbage and turnip grown for seed, fireweed, peppermint, black locust, snowberry, vetch, and several others along with honey from alfalfa and the clovers.

Since there are so many plants which are important locally, it will be to the reader's advantage to ascertain whether a bulletin on honey plants of his state may be available from his college of agriculture. "American Honey Plants," by Frank C. Pellett, is the title of a volume of more than 400 large pages devoted to a discussion of honey plants for all America.

HONEYDEW

At times when nectar is unavailable the bees gather a sweet which is excreted by aphids feeding on the leaves of forest trees. This honeydew is usually of poor quality and of little value. It is bad for the bees when left in the hives for winter stores where winter weather is too severe to permit the bees frequent flight. At times when aphids are abundant they eject honeydew in such quantity as to cover leaves on lower branches with the sticky substance. The bees gather this material and seal it in their combs the same as honey. When mixed with good honey it spoils the quality of the whole output.

BITTER HONEY

Fortunately the areas where unpalatable honey is harvested are relatively small. In the Ozark region of Arkansas and Missouri and eastward in Kentucky and Tennessee the bitterweed yields an abundance of bright sparkling honey which is as bitter as quinine. It is important that the beekeeper remove such honey from the hives before the harvest of good honey. It can be fed back to the bees for winter stores after the other is taken from the hives.

	Earliest Pollen Nectar Flowers	Late Spring Build-up Flowers	Main Honeyflows	Minor Sources	Fall Flowers
Northeast	skunk cabbage willow maples	fruit bloom dandelion black locust	clovers goldenrod fireweed buckwheat	sumac buttonbush Canada thistle milkweed purple loosestrife	asters wild carrot goldenrod Spanish needle
Southeast and Gulf Coast	alder yellow jasmine wild plum	fruit bloom dandelion tulip tree chinaberry	clovers mesquite sourwood aster persimmon gallberry	holly sweet clover tupelo black locust basswood cotton	golden aster asters goldenrod summer farewell
Texas and Southwest	mistletoe creosote bush pinkmint	fruit bloom dandelion arrow wood	cotton mesquite catsclaw huajillo	horsemint sweet clover alfalfa	rabbit brush asters Spanish needles
Central West	maple willow elm (pollen)	dandelion fruit bloom Virginia waterleaf	white clover sweet clover heartsease	basswood Spanish needle coralberry	wild sunflowers asters bonesets
Plains States	prairie crocus willows wild plums	dandelion box elder cottonwood black locust	sweet clover alfalfa	gumweed Rocky Mt. bee plant	goldenrods wild sunflowers ironweed
Northwest	vine maple willows	orchard fruits dandelion	alfalfa sweet clover fireweed	dogbane turnip cabbage black locust cascara snowberry	goldenrods rabbit brush
California	willows dandelion	cantaloupe onion fruit bloom	orange alfalfa sage clovers star thistle	eucalyptus lima beans mustard manzanita blue curls	rabbit brush Spanish needles
Florida	wild pennroyal Spanish needle	citrus blackberry chinaberry	gallberry tupelo palmetto black mangrove	ti-ti manchineel purple flower mint citrus	summer farewell goldenrod asters

CHAPTER VII

THE HONEY BEE AND POLLINATION

A LTHOUGH the honey bee has been famous for centuries for her labor in the production of honey and beeswax, she is of far greater service to mankind in the distribution of pollen.

A large number of plants are fruitful only when pollen from the flowers of one plant is brought to the flowers of another to insure fertilization. Under present day conditions the honey bee is by far the most efficient agent in this pollen distribution, commonly spoken of as pollination.

Many of our most important food plants are dependent upon the bees for this service. The grains are largely wind or self pollinated and do not depend upon insects; but most of the fruits, many of the legumes, including the clovers and alfalfa, and such garden vegetables as cucumbers, melons, and cabbage require the help of insects to insure that fruit will be formed or seed set.

Fig. 1. A pollen-covered bee in flight. The sticky pollen grains adhere to the body hairs and are also carried in the pollen-baskets on the legs.

In a recent publication of the United States Department of Agriculture, Circular E-584, is given a list of numerous plants which are important in present day agriculture, all of which either depend upon the honey bee for pollination or are helped by her visits.

In the days of the self-sufficient farm we knew little of the problems which have developed along with the present day specialization. Every farm raised a little of a great many different things and the family provided a home market for about sixty per cent of the farm output. Now a farmer will devote himself to the production of two or three crops and often to only one. The large areas devoted to special crops provide ideal conditions for the spread of insect pests with the necessity of heavy expense of time and money in the effort to control them. The poisons used in their control often destroy the wild insects which serve a useful purpose in the distribution of pollen. With the disappearance of the many wild bees, ranging in size from smaller than a housefly to the bumblebee, we have come to be dependent upon the honey bee which is the only insect whose numbers can be controlled and which can be moved to the spot where her services are most needed.

In neighborhoods where there are large areas of wasteland and where little spraying is done, there may be plenty of wild insects still present to serve the need of small acreages. However, it is common practice at present to grow special crops in such a large acreage as to require a far greater number of insects than are present in the wild.

LOSS FOR LACK OF POLLINATION

It often happens that unprofitable crops will result from lack of pollination when growing conditions are favorable and the grower is at a loss to understand his failure. In many localities where farmers once harvested five to eight bushels of red clover seed per acre, they now get less than one bushel and seed production is no longer profitable. This change came with the disappearance of the bumblebees which were once plentiful in the area.

The honey bee is not as efficient in the pollination of red clover as the bumblebee but when enough honey bees are present they do provide the necessary service. Since the corolla tubes of the red clover are so deep that it is difficult for the honey bees to get the nectar they are likely to visit other flowers in preference if any are open. The remedy is to bring in so many honey bees that other pasture will not be sufficient and they must visit the red clover to provide their needs.

It would be difficult to estimate the extent of losses to agriculture from lack of pollination. The difference in yields in fields or orchards after the honey bees have been brought in, has shown that such losses run into hundreds of millions of dollars. The loss of soil fertility through continuous stirring of fields that should be sown to meadow or pasture is even greater and this loss is permanent and cannot be replaced.

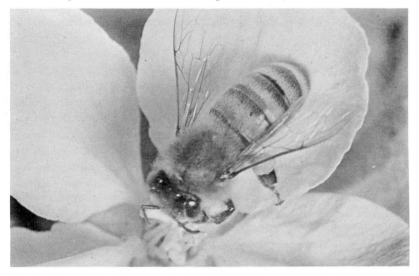

Fig. 2. A bee visits a blossom and grains of pollen become lodged on the body hairs. Some of these grains may be brushed off as she visits the next blossom. (Photo, W. P. Nye)

Fig. 3 The result of cooperation and harmony in Nature. Man has learned to help this process along by renting bees for orchard and crop pollination. (Photo, J.C. Allen and Son)

COMPETITION FOR ATTENTION OF BEES

In any program of pollination it is necessary to bear in mind that the bees are likely to show preference for flowers with nectar of the highest sugar content. Thus, we cannot be sure when we bring bees to our orchards or gardens that they will visit the flowers which we are so anxious for them to serve.

The nectar from some flowers is rich in sugar while others may have a low sugar content. Since the bees are constant to one kind of flower while it is in bloom and visit only that species while its flowers are open, it is important for the grower that they start visiting the crop to be pollinated when its flowers open. This constancy greatly increases their value as pollination agents since they carry only pollen from apples when they are gathering from apples or from dandelions when that is the source of their nectar.

Some red clover seed growers have decided that the honey bee is a poor pollinator for red clover when the insects flew over their fields to visit sweet clover over the hill. Likewise, the owners of pear orchards are often disappointed when the bees fail to notice their fruit trees an seek apple orchards farther away. In turn the apple grower may be disappointed when the bees find dandelions in nearby pastures more attractive.

The remedy, as already stated, is to make sure there are enough honey bees in the neighborhood to insure visits to the flowers with low sugar content when that is necessary.

There is great variation in the number of bees necessary to insure pollination of the various crops. Much will depend upon the acreage within flying range, the length of the flowering period of the plants to be visited and the number of wild bees already in the neighborhood. The weather at flowering time is also important. At times when weather is chilly and cloudy with brief periods of time suitable for bee flight far more bees will be needed than when weather is balmy and bees can fly from early to late.

Under favorable conditions one strong colony of bees per acre may be sufficient where five would be needed if the weather is bad at time of bloom. There are so many unpredictable factors that it is good policy to bring in more bees than are thought to be necessary. Extra bees bring added assurance that the crop will be pollinated.

Every crop offers problems peculiar to itself and every neighborhood has a different set of conditions. It should be borne in mind however as stated earlier, that only strong colonies offer efficient pollination service. The beekeeper who rents bees for pollination should be prepared to give service on this basis to insure that his customer will get the value of his money.

CHAPTER VIII

SWARMING

A swarm of bees tumbling out of a hive in pulsating bursts is one of the most fascinating sights of Nature. It can also be one of the most frustrating moments for the beginning beekeeper. He has purchased the bees and equipment, hived the package and carefully nursed them through to the beginning of the honeyflow. Now it would seem as through, for reasons known only to themselves, they have abandoned their jobs, eaten up the precious profits, and absconded with the business to go on a rather mindless and carefree holiday at the beekeeper's expense. In fact, one type of swarming is known as "absconding," a rather maverick, cross-country jaunt, usually resulting in total loss.

Swarming is one of Nature's ways of dividing colonies to create new ones. In the wild, swarming propagates the species and is responsible for the presence of bees on earth today. In a man-made environment, swarming is often a symptom of problems within the hive, and the opposite of being mindless, is a planned and calculated exodus on the part of the bees. The bees are attempting to solve problems of overcrowding, improper ventilation, starvation or other problems by swarming in much the same way as people move from crowded cities to form suburbs and housing developments.

Swarming usually takes place in May or June in the North. For some days before swarms issue, the bees may be seen clustering at the entrance of their hive, though some swarms come out when there are no indications. When honey is abundant and bees are plentiful, look for swarms to issue at almost anytime from the hours of ten a.m. to three p.m. for first swarms; for second and third swarms from seven a.m. until four p.m.

THE CAUSES, SIGNS AND PREVENTION OF SWARMING

The major causes of swarming are: overcrowding, improper ventilation, and supersedure impulses. All of the above either create or are the result of improper conditions within the hive. Although some are overlapping, each has its individual causes, is detectable by its characteristic signs, and is often preventable by various management techniques. Since swarming through any of these means reduces the worker population and therefore, reduces the amount of honey crop, it should always be prevented if possible. If the beekeeper fails to diagnose these problems in time to take corrective measures, the bees will solve them in the only way Nature has taught them—swarm!

Overcrowding If bees are crowded for room they will at once make preparations for swarming. Just before the main honeyflow begins there is usually a lull in activity in the colony. By this time the colony has reached its numerical peak and without abundant flowers in bloom there are thousands of bees with little or nothing to do. Even big colonies with the best of queens are frequently unable to stand the congestion of a crowded brood nest. This congestion throws the colony out of balance in population and queen cells are started in preparation for swarming.

Signs of overcrowding are relatively easy to spot. There will be bees hanging onto the front of the hive, about the entrance, and clustering about the corners of the hive. A congested brood nest is simply one that is out of free space, its cells either full of sealed brood, honey, or pollen. Another sign of overcrowding may be burr comb, irregular bits of comb constructed in any available space, on tops of frames, in corners, or anywhere there is room to construct a few cells to use for brood (egglaying) or storage of honey and pollen.

Fig. 1. This colony has outgrown its hive. Unless additional space is provided for brood and storage of honey and pollen, the colony will swarm.

Swarm prevention under these circumstances will certainly call for keeping the brood nest as free as possible at all times from any condition leading to congestion. There should be plenty of worker combs containing empty cells for egg laying and with a minimum of honey and pollen in them. If there is room at the sides of those combs occupied with brood, add mare empty combs so that the queen may use them, or room may be provided for the empty combs by removing those occupied with storage of honey and pollen. Thus the queen is induced to expand her brood area to the sides.

The brood nest may also be expanded upward by supplying another brood chamber on top of the congested one. Supers can also be added, supplying an area for the storage of nectar and honey away from the brood area. This is the basic idea of a free brood nest by the provision of areas for definite use.

Fig. 2. This hive is not likely to swarm for some very good reasons: 1. It uses two brood chambers to eliminate overcrowding. 2. It has been given a honey super which will occupy the bees with drawing comb for storage of honey. 3. It has a full entrance and light shade which will aid in ventilation, and 4. It has the attention of a beekeeper who understands the needs of the bees.

Adding the first honey super will give the bees plenty to do for they must draw the foundation out into comb before they can store honey. Adding this first super can be a very effective swarm control measure if properly timed and executed. Remember that drawing foundation into comb is an expensive process for the bees. **Be sure the bees have sugar syrup at this time if natural sources of nectar are not yielding.**

Improper ventilation is a condition often found with overcrowding. The presence of so many bees in the hive creates excessive heat. A sign of this condition may be bees fanning at the entrance of the hive, attempting to lower temperatures inside by setting up a current of air moving through the hive. Bees may also be clustering about the outside of the hive, unable to stand the excessive heat inside.

As the colony grows and increases in size, give it a larger entrance opening. The entrance reducer supplied with the original beekeeping kit has two separate openings. When the package is first hived it is best to use the small opening, but by the end of the fourth week the large opening should be used. As the honeyflow approaches the entrance block should he removed entirely and the bees given use of the full entrance to the hive. This should provide adequate ventilation, but if the weather is extremely warm, it would be well to place two small blocks of wood between the hive body and the bottom board. This will raise the hive and tilt it backwards and allow ventilation along the sides as well as increased ventilation in front.

Supersedure of the queen is Nature's way of replacing a failing queen. The worker bees take matters in their own hands and construct queen cells in an attempt to replace their failing mother.

Supersedure in package bees occasionally presents a problem. Some times it is caused by too frequent handling of the colony. Such supersedure in package bees usually occurs about three weeks after the package has been installed. Certain queens also are inferior to others in the quality and quantity of their brood, a factor which contributes to queen supersedure.

The beginner is apt to be confused as to the difference between queen cells constructed for the purpose of supersedure and those constructed for the purpose of swarming. Actually, there are many oldtime beekeepers who are unable to distinguish the difference.

Simply stated, supersedure cells are few in number— usually only three or four are constructed—are of the same age, and therefore, in the same state of development, and usually are larger and more copiously supplied with royal jelly than are the swarm cells. They commonly are constructed outward from the surface of the comb or in some depression along the side of the comb. Swarm cells are numerous usually ten or more being constructed. They are of varying size and one or two being started each day over a period of a week or more, and are primarily constructed along the bottom edge of the combs.

Since supersedure may occur at the height of the prosperous period before the honeyflow, the colony may dovetail their swarming with their efforts to secure a new queen by supersedure.

Fig. 3. Supersedure, queen cells. Notice that these cells are all of the same approximate age. Supersedure cells tend to be large and lavishly supplied with royal jelly. Notice how these cells were built next to the damaged area on the face of the comb. Some authorities tend to the belief that a queen raised under the supersedure impulse cannot be surpassed for quality. As a result of supersedure it occasionally happens that two queens, mother and daughter, inhabit the same hive.

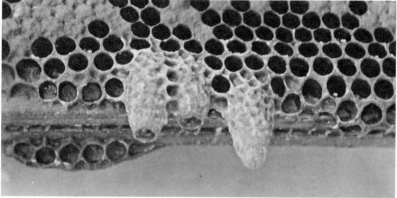

Fig. 4. Queen cells built under the swarming impulses. Notice that in this case there are three cells of varying ages. The cell on the right is already sealed and in the pupal stage. The queen cell on the extreme left is just approaching the sealing stage, while the center queen cell is still in the mid-larval stage and will be the last cell to be sealed.

If there is evidence of swarming, all of the queen cells should be destroyed. However, if there are only a few cells and evidence of an attempted supersedure it would be best to leave two sealed cells in widely separated parts of the hive. When the bees are trying to supersede the queen and all cells are destroyed there is the possibility that the colony will become queenless in a short time. Usually the worker bees do not attempt to supersede the queen unless there is evidence that she is failing in her allotted role of egg-laying.

Fig. 5. Good brood—the result of an excellent queen

A queen's quality may readily be recognized by the way in which she deposits her eggs in the cells. A good queen will lay in a compact area so that the sealed brood presents a solid mass of wax cappings. A poor queen usually lays spottily with eggs scattered throughout the comb leaving many cells empty. Occasionally a queen will become a drone-layer—i.e., she will lay unfertilized eggs which will develop into nothing but drones. Such a queen should be replaced as quickly as possible. Order a new queen, and when she arrives kill the old one and place the new queen in the colony. (Instructions for introduction of a queen are included in shipment of the queen cage.)

Fig. 6. Poor brood—the result of a drone-laying queen. She should be replaced as quickly as possible with a young, hybrid queen.

SWARMING

After whirling a few minutes in the air, the mass of bees will cluster on the branch of some tree or bush close by—generally one that is shaded from the sun. They should be hived as soon as the cluster is formed, else they may leave for the woods, or if another colony should cast a swarm while the first is clustered, the two would probably unite.

Should the queen fail to join the swarm, because of having her wings clipped, or for any other reason, the swarm will return to the hive as soon as the bees are aware of her absence. As they are gorged with honey, the bees, when swarming, may be handled without fear of stings.

The queen has very little to do with swarming preparations. If a second swarm is wanted by the worker bees, they will prevent the first queen from destroying the other queen cells which are sometimes very numerous even on small pieces of comb. She would be sure to destroy the other queens if given a chance. If restrained, she will show her irritation by piping, and this piping is answered by the other queens which are kept like prisoners in their cells. The second swarm then issues within two or three days.

After the departure of this swarm and the emerging of the second queen, if her piping is also answered by a third queen, a third swarm may also issue. If the desire to swarm is satisfied after the departure of the first swarm, all the queen cells will be destroyed by the first young queen that emerges. The worker bees will often help her in this task; seemingly anxious to return to the normal state of one queen per colony.

HIVING THE SWARM

It is easy to hive a swarm if the swarm cluster is within reach. A hive should be in readiness, with empty brood combs and with a full entrance in front. The limb on which the swarm is hanging may be cut off and the swarm carried to the hive and shaken before it. The queen, if she can be located, may be directed toward the hive. She will enter it eagerly. for she loves darkness. The bees will crawl into the hive, and finding the queen, will be satisfied to stay. When the bees are inside, the hive should be placed where it is to remain; a shaded position is the best. Frames of comb foundation should be given to the swarm, along with a few frames of drawn comb to support the cluster.

If the bees have clustered on a branch which is too valuable to be cut down, a basket, box, or swarm sack, into which to shake or brush the bees, will be essential. If they are on a wall or fence or on the trunk of a tree, it is possible to brush them into a basket and to proceed to hive them.

Sometimes it is impossible to brush the swarm into a box, or perhaps the number which may be gathered together at one time is so insignificant that the bees take wing at once to return to the cluster. Then a comb of brood

Fig. 7. Fortunately, this swarm lit among the low-hanging branches of a nearby tree. The beekeeper has cut the branch, carried it to the hive, and is preparing to shake the branch in front of the hive. (Photo, courtesy of L. E. Strader)

from another hive, or even a frame of dry comb placed over the swarm will soon be recognized by the bees as of possible use to them, and they will readily cluster upon it; or they may be gently smoked to direct them towards it. Care must be taken to not frighten them away.

Perhaps two swarms have taken wing at the same time and have clustered together. They may be evenly divided by placing two hives on the ground and directing the bees equally to both, especially if the queens are found and caged and placed at the entrance of the hives. Bees may be scooped or shaken without much trouble when the swarm is gathered, and they are easily directed to one hive or another if properly handled.

A frame of brood placed in the new hive will be of much advantage to the bees. It will prevent the swarm from leaving the hive; and should the queen be lost, it will give them the means of rearing another. If the other frames have been filled with comb foundation, the bees will soon be in good condition and perfectly at home in their new quarters.

CHAPTER IX

THE HONEYFLOW AND REMOVAL OF THE CROP

WITH REASONABLE care on the part of the beekeeper and a satisfactory buildup season, colonies should be full of bees at the approach of the main honeyflow. Colonies should have ample brood with sufficient room for the egg laying of the queen; the bees should be without intentions to swarm and the morale of the colony should be such that only abnormally cold and rainy weather will hinder the gathering of surplus.

Every beginner in beekeeping is confronted with the problem of what kind of honey to produce. Bees gather nectar, subject it to a chemical process which results in unripe honey, and store it in cells. Later, it is capped. A frame of honey consists of thousands of these capped cells, constructed side by side and back to back. This is the way of the bee. It is up to the beekeeper to alter this basic procedure to produce the particular product he desires.

Extracted honey is the liquid honey removed from the comb. The comb is then returned to the bees for refilling. Comb honey is marketed in the comb which is consumed as part of the product. Comb honey is produced in two forms: section comb honey and bulk comb honey, which is later processed into chunk honey or cut comb honey. No matter what kind of honey is produced, the beekeeper should have supplied each colony with at least two supers in which to store the honey.

MANAGEMENT FOR BULK COMB HONEY

Since the beginner usually has only one or two colonies of bees, it is recommended that bulk comb honey be produced. Production of bulk comb honey does not require the high degree of professional "know-how" that is needed to produce section comb honey, nor does it require the additional equipment needed to produce extracted honey. Later, if the beginner wants to expand and go into extracted honey production, the bulk comb honey supers may easily be converted to extracted honey supers.

Bulk comb honey usually is produced in shallow supers similar to those used in producing extracted honey. The frames are filled with a high quality of thin surplus or bulk comb foundation, that is made especially for that purpose. The resulting combs of honey are as delicate as that of section comb honey. The only difference between bulk comb honey and section comb honey is the size of the combs—the slabs of bulk comb honey weighing from 3-1/2 to 4 pounds, while the sections of comb honey weigh from 13 to 16 ounces.

The first bulk comb honey super should be put on as soon as the main honeyflow has started. The second super is placed on top when the first one is about half filled with honey. When the bees begin to draw the foundation

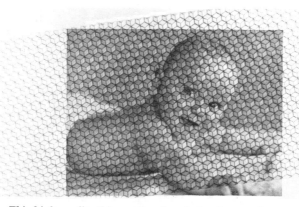

Fig. 1. This high quality thin surplus foundation is manufactured so thin that a picture may be clearly seen through it.

in the second super it should be placed down next to the brood nest and a third super added on top. Additional supers are added and manipulated in the same manner. This method of super manipulation keeps the bees spread throughout the supers. They have work ahead of them all the time and there is little tendency on their part to clog the brood nest with honey. Toward the close of the flow care should be taken not to put on too many supers. This will force the bees to fill and finish the ones that they have so that there will be a minimum of partially filled and sealed combs of honey.

Sometimes the bees will fill with honey and seal the combs in the center of the super and leave the outside combs untouched. When this happens you should reverse their position by moving the outside frames to the center of the super and the center frames to the sides of the super. This will force the bees to draw out and fill all of the combs.

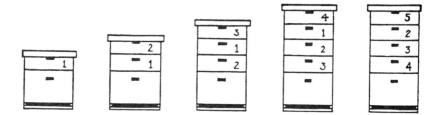

Fig. 2. Diagram showing the method of supering. The supers are numbered in the order in which they are given to the bees.

The supers of bulk comb honey are removed as soon as the combs are fully capped. If they are left on too long they will become soiled and travel—stained by the numerous bees that must pass over them.

If the producer is to use the honey produced for his own consumption it may be kept in the large frames and cut out whenever more is needed for the table or to give to friends. If the honey is to be sold there are various ways of preparing bulk comb honey for the market.

Fig. 3. Chunk honey packed in a jar is a fancy honey. The beekeeper should always be sure that the finished product is neat, clean, and attractive. The comb should be white and free from pollen or stains and the liquid honey should be clean and clear.

CHUNK HONEY

When packing chunk honey, the bulk comb honey is cut into rectangles which fit into the mouth of a glass jar. The remaining space in the jar is then filled with liquid honey which has been previously heated to 150 degrees F and then cooled to room temperature before it is poured over the honey chunks in the jar. As soon as the jar is filled the lid should be placed on tightly.

Chunk honey should not be stored for too long a time since it is apt to granulate. Jars of honey that have granulated may be liquefied by placing in a warm oven. (The temperature should be kept below 145 degrees F. which is the melting point of beeswax.)

CUT COMB HONEY

Cut comb honey is bulk comb honey cut up into varying sizes or pieces. Cut comb usually varies in size from a small 2-ounce individual serving to larger ones weighing as much as a pound.

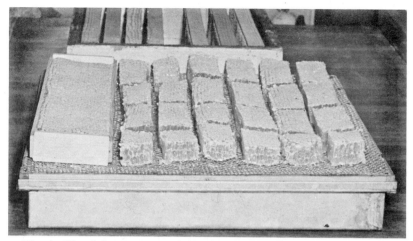

Fig. 4. Chunk honey, cut and draining before packing for the market.

The pieces of cut comb honey are drained on a screen in a warm room for 24 hours. If an extractor is available they may be placed in a screen basket in the extractor and the liquid honey along the cut edges will be thrown off by centrifugal force. The well-drained sections may be placed in specially-designed plastic cut comb boxes. The finished product resembles section comb honey.

Fig. 5 Cut comb honey attractively packaged in leak-proof, clear plastic containers finds a ready market.

MANAGEMENT FOR EXTRACTED HONEY

Because of the initial expense involved, the beginner is not urged to produce extracted honey. Extra equipment required to produce extracted honey includes such items as an extractor and an uncapping knife. In a large apiary, or from the standpoint of commercial beekeeping, extracted honey is the cheapest to produce, since one man may care for more colonies in less time than with the two previously mentioned systems.

The early management of colonies to be used in extracted honey production is no different than that of section and bulk comb honey. No matter what type of honey is being produced, the colonies involved should be at the peak of their strength when the honeyflow begins.

The foundation used in the extracting super frames differs from section and bulk comb foundation in that it is made of a somewhat darker and heavier wax. It is also wired to prevent the comb from breaking in the centrifugal extractor. In producing extracted honey the same combs may be used year after year, while in bulk and section comb honey the supers are filled with new foundation each year.

Supers of drawn comb are added to the top of the hive in extracted honey production As the top super is filled with honey and the bees have

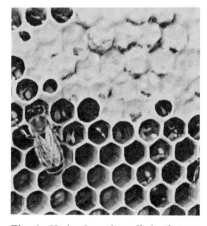

Fig. 6. *Notice how the cells in the top of the picture are white with new wax and how the bees are sealing the full cells of ripe honey.*

Fig. 7. *Jars of extracted honey after drawing from the storage tank. With the addition of a colorful label, these jars will make an attractive display.*

started to whiten the comb and seal the cells, another super is added on top of the one already on the hive. The beginner, starting out with all new equipment and supers of foundation instead of comb, should follow the same supering procedure and manipulation as given for bulk comb honey pro-

duction. Usually four supers per hive are enough. If the honey crop is larger, and more supers are needed, the full supers may be taken off and extracted and the supers with empty combs placed back on top of the colony.

In extracting the honey from the combs the first step is to remove the
wax seal over the top of the cells. This is accomplished by means of an
uncapping knife. The knife slices the wax seal from the cells so that the
honey may be thrown out by centrifugal force in an extractor. For best
results the honey should pass from the extractor, through strainers, and into
a storage tank. After settling for 24 hours, the honey may be drawn from the
storage tank into cans or bottles to be used by the producer or sold on the
market.

REMOVING THE HONEY CROP—COMB HONEY

Comb honey is removed by means of an escape board. The escape
board is your inner cover with a bee escape placed in the hole in the center.
The bee escape allows the bees to go down out of the super but prevents
them from returning.

The super to be removed is first placed on top of the colony and given
a few quick puffs of smoke to start the bees running down. It is then lifted
off the colony and given a few vigorous shakes in front of the hive. The
inner cover with the bee escape in place is then placed on top of the hive and
the super to be removed placed on top of the super, taking care that there is
no crack big enough to allow bees an entrance to the super. Usually the bees
have completely left the super within 24 hours.

BULK COMB HONEY AND EXTRACTED HONEY

Bulk comb and extracted honey may be removed by the same process
as that described for comb honey. Another way is to take the full supers of
honey and brush the bees from each individual comb with a bee brush. For
the beginner, either of these two processes will be satisfactory.

Many large commercial beekeepers find it to their advantage to use a
repellent in removing supers of extracted or bulk comb honey. Properly
used, a good repellent saves much time and labor in the commercial apiary. Another good investment for the larger apiary is the bee blower. The
bees are removed quickly and safely by directing a stream of air against
them.

ROBBING

During any period of the year when there is no nectar coming into the
hives the bees are apt to rob. The bees from a strong colony will attempt to
get into another which is weak in numbers and obtain the honey. Strong
colonies in a yard, if given an opportunity, will steal all of the stores of

Fig. 8. When the supers are removed and placed on the hive cover, a robber cloth spread over them protects them from robber bees. Next, a second cloth covers half the brood and hangs over the side; a third covers the other combs. Here one cloth is being pulled back so that a comb can be removed. The cloth is replaced over the exposed area before checking the comb.

honey of the weaker colony. This usually results in the death of the colony which has been robbed. Weak colonies should have the hive entrance reduced to a minimum in order to discourage robbers. The possibility of robbing is likely just after the honeyflow is over. Because of this, particular care should be taken when removing supers of honey not to get robbing started.

One good way to prevent robbing is the use of cloths dipped in a mixture of water and kerosene and the excess liquid wrung out. The damp cloths are then used to cover the hive and supers while working a colony. Be sure the cloths are damp and not wet enough to drip water and kerosene on the honey.

Worker bees which act as guards are present at the entrance of hives at all times. They detect the presence of and repel bees which do not originate from their own colony. When worker bees are seen flying around the entrance to the hives and are being repelled by the guards it is an indication that the bees are trying to rob. Whenever there is an indication that the bees are trying to rob it would be best not to smoke the entrance of a colony before working it. The smoke will disrupt the guard bees and allow robbers to gain an entrance to the colony.

EXAMINATION FOR DISEASE

Periodical examination of the colony and its brood should be made to protect against brood diseases. This subject is treated in Chapter XIV. Suffice to say that while diseases, when they appear, may cause great damage to the colony, the use of disease resistant bees and the use of Terramycin® and fumidil B in the colony feedings have done a great deal to keep down the incidence of disease.

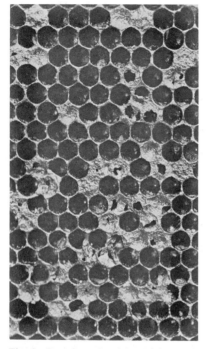

Fig. 9. Brood comb, with an advanced case of American foulbrood. Notice the perforated and sunken cappings on the cells and the scattered appearance of the brood. many of the open cells, from which cappings have been removed by the bees, contain scales. When the larva has died with the foulbrood, it dries down into this scale.

Fig. 10. As decomposition of the larva progresses, the dead larva sinks down in the cell and the color changes to dark brown. During this stage the larva has a ropiness to it. With a match or toothpick the remains may be drawn out like thick glue into fine threads. Typical dark brown foulbrood scales form on complete drying.

---

*Terramycin® is a product of Pfizer & Co. inc.

CHAPTER X

HONEY AND BEES WAX

HONEY AND beeswax are the two principal products of the honey bee colony. The pollen gathered by the bees from the flowers is mixed with honey or sugar syrup within the hive to feed the bees. Some experiments have been made which suggest that pollen may have some possibilities as a human food.

The propolis gathered by the bees (similar to glue) and which the bees use to close up holes and cracks in the hive as well as to cover objects within the hive which the bees cannot remove, has been used at times as a furniture glue and polish. The acid of bee stings is being marketed as an inoculation for certain types of arthritis.

HONEY

The honey bee visits the flowers which secrete a sweet liquid called nectar. This nectar, waterlike in consistency, is sipped from the blossoms by the bee and carried to the beehive, in its honey stomach. While some enzymes may be added on the trip and some thickening may take place, the raw nectar goes into the cells in almost the same condition as it was when withdrawn from the flowers. Here, within the hive, the house bees evaporate the nectar down to a thick consistency which is the commercial honey of today.

Fig. 1. Bees collecting sugar syrup from a container. Notice how the proboscis is used to suck up liquid food. The bees sip nectar from flowers in much the same manner. (Photo, J. B. Free)

Such honey contains from 15 to 20 per cent water (usually about 18 per cent), 40 per cent levulose or fruit sugar, 34 per cent of dextrose or grape sugar, and 2 per cent of sucrose or cane sugar. There is besides, in the honey, a small percentage of dextrine, ash, acids and minerals and a small quantity of undetermined materials. When honey contains 20 or more per cent water it is unripe and under some conditions may ferment. For that reason the beekeeper is urged to leave the honey on the hive until the bees have thoroughly ripened it. Sealed honey, except under extreme conditions, is ripe. The bees often ripen the honey as it is gathered. It is only under circumstances of heavy honeyflow or high humidity that they have difficulty in keeping up with the field-gathering bees. At the end of the flow, all honey, sealed or unsealed, is safe to remove. Ripe liquid or extracted honey weighs 11-1/2 to 12 pounds to the gallon.

Honey is classified according to the source from which it is collected. Thus, we have a wide variety of honeys depending on the major source from which they were gathered, whether it be black sage, orange, sourwood, clover, alfalfa, buckwheat, milkweed, etc. Similarly, in most localities where there is a variety of plants, we are apt to have a mixture of honey from several flowers, the honey taking its color and flavor from the combination of flowers represented. In our chapter on "Sources of Bee Pasture" we give a chart with names of the principal honey plants in various sections of the country.

Honey also is classified according to its color and flavor. It may either be white, amber, dark amber, or straw colored, and may be either mild, or more pronounced in flavor. Usually the light-colored honeys are milder in flavor, while the strong-flavored ones generally are darker in color. The darker honeys have somewhat higher mineral content and are preferred by some for this reason. Usually the kind of honey we have been used to in our neighborhood, is the honey which appeals to us. The New Yorkers and Pennsylvanians like their buckwheat. The northern sections generally produce the clovers which are preferred there, while the Southerner prefers sourwood, or mesquite, or tupelo, or gallberry.

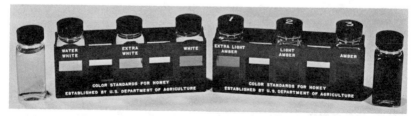

Fig. 2. The U.S.D.A. color comparator contains bottles filled with materials resembling different classifications of honey. The actual honey is compared with the samples until a classification is reached.

There are few honeys which are objectionable. The bitterweed of the South produces a bitter honey which should be left on the hives for winter food and spring build-up. But even the bitterweed loses much of its bitterness when off the hive and exposed to the air.

Most honeys granulate within a few months after harvesting, the ones with the least water content being the most rapid to granulate. Thus. the honeys of the arid regions are much more apt to granulate rapidly (candy or sugaring) than those where the humidity is greater. Granulation of honey is nearly always a sure proof of the purity of honey, though nowadays the public is protected from adulterants by the pure food laws. Tupelo honey, to the contrary, seldom granulates because of its relatively high content of levulose sugar and a minimum of dextrose. It finds a ready special market.

A honey which has been heated as a preventive of granulation will be slow to granulate again. So, the bottlers and packers of liquid or extracted honey heat the honey as rapidly as possible to a maximum of 150 degrees, being careful not to injure the honey by overheating or scorching. This is usually done in steam or water jacketed tanks with a very careful control of the temperature. The honey is then drawn off quickly into retail receptacles and sealed at once while hot. This helps retard further granulation also. But care must be taken to allow the honey to cool as rapidly as possible after it is sealed, otherwise, if stacked in a close pile, the retardation might definitely darken the honey and impair its flavor.

Similarly when honey is kept over a long period, every effort should be made to keep it in as cool and as dry a place as possible. Temperatures of 80 or 90 degrees or more over a long period will cause the honey to get darker, and as honey is hygroscopic (capable of absorbing moisture), a damp storage place may cause it to become thinner and to ferment. So honey may show tendencies to ferment either from having been harvested from the hive when yet too high in moisture content, or it may ferment if stored where the air is exceedingly moist. When, a honey has once soured, it will be hard to return it to its original flavor. though prompt heating of a slightly soured honey may remove most of the difficulty. Hopelessly sour honey can be fed back to the bees after heating.

As already explained, honey is usually sold either in comb honey sections, as wrapped cut comb honey, as chunk comb honey which is a mixture of both comb and extracted honey, or as extracted honey. The bulk of American honeys moves as liquid or extracted honey. Granulated honey should have more of a market than it has. Lately there has been developed a process by which the honey may be aided in its granulation by the addition of very fine grains of candied honey. This process called the Dyce process makes a very fine smooth granulated honey called creamed honey, which is reaching a popular demand.

HONEY USES

We think of honey usually as a table spread, to use on bread or pancakes, or biscuits. This is possibly its chief use. Its use in cooking, however, is very considerable, both in pastries, in canning, in mill; drinks, in syrups and desserts and as an adjunct to frostings and dressings.

BEESWAX

Beeswax is the second most important product of the beehive. Produced by the honey bees themselves in the hive, it is extruded between the segments of the underside of the abdomen of the bee while she is gorged with honey. These small slabs of beeswax are formed either into comb for the bees' brood, their combs for surplus honey, or for sealing the combs when they have been filled with honey. The beekeeper obtains beeswax by melting up and pressing the wax from old discarded empty brood or super combs; or from cappings which have been removed from sealed combs of honey in extracted honey production.

Beeswax was one of the earliest of waxes, being used in the form of candles for lighting. This remains at least the second largest use of beeswax today. Several church faiths use beeswax to some extent. Earlier the Roman Catholic Church required that pure beeswax be used on the altar for the ser-

Fig. 3. Pure, natural honey about to be put to a very good use. (USDA photo)

Fig. 4. The keeping qualities of honey make it a natural ingredient for breads, rolls, cookies, pastries, and pies. (Photo, Honey Advisory Board, Calif.)

vice of the Mass and the Benediction of the Blessed Sacrament. As the numbers in this faith grew, however, there was not enough beeswax to serve the purpose and the regulations were modified to require candles which were at least 51 per cent beeswax.

Probably the largest user of beeswax today is the cosmetic industry, since the emulsifying agent of nearly all of our modern cold creams is dainty, pure, white bleached beeswax, being used as well in ointments, lipsticks, pomades and rouges.

The beekeeper himself is the third largest user of beeswax in the form of comb foundation which he gives to his bees as the base for their combs. There are some 70 or more commercial uses of beeswax today. Some 200 million pounds of honey and four to six million pounds of beeswax are produced yearly in the United States.

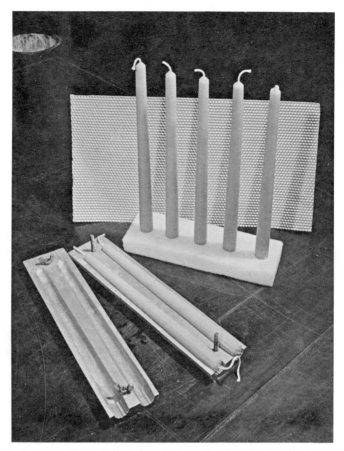

Fig. 5. Beeswax candles and comb foundation to be given back to the bees are two of the many uses for beeswax. (Photo, Mich. St. Univ.)

CHAPTER XI

FALL MANAGEMENT AND WINTERING OF BEES

THE END of the midseason flow is usually followed by a period when there is little or no nectar coming into the colony. This period of relative inactivity is followed in many localities by the fall flow-usually coming from such flowers as heartsease, commonly called smartweed, aster, goldenrod, Spanish needle, white boneset, wild cucumber, and climbing milkweed.

The beginning of the fall flow is one of the times during the year when the influence of the queen is felt greatly. An inferior queen at this time will lead to a colony that is weak in worker population and incapable of wintering successfully. If there is the slightest indication of a poor queen, the colony should be requeened during the fall flow.

In an area which has little or no fall flow, or if queens are available during the fall flow, colonies may be safely requeened after cold weather has set in and brood rearing has ceased. Go into the colony to be requeened on a cold day in late September or early October, remove the old queen, and

Fig. 1. Boneset is an autumn bloomer and the honey is usually mixed with that of heartsease, asters, goldenrod, Spanish needles and other plants blooming at the same time.

Fig. 2. Common goldenrod is one of the most widely distributed native plants, growing in fields, open woodlands and hedgerows. The odor is faint, but the nectar is clearly visible in the flowers.

place the new queen in the mailing cage in the center of the cluster of bees. Make sure that the cardboard or cork plug has been removed from the candy end of the queen cage.

Care must be taken during the fall flow not to place too many supers on the colony. Additional supers should be added only when the one on the

Fig. 3. A colony ready for winter should have a rim of honey along the tops of the brood combs as well as a full super of honey.

colony is practically full of honey. This prevents the bees from spreading the incoming nectar throughout more supers than they will fill. When the fall flow is over—usually after the first killing frost the supers may be removed from the hives.

During the removal of the fall crop, one super completely filled with honey should be left on the colony. In addition to this super, the colony should have from twenty to thirty pounds of available sealed honey in the brood nest. This honey in the brood nest will usually be found along the tops of the brood combs. The honey in the super, plus the honey in the brood nest, should give the colony from sixty to seventy pounds of honey for their winter and early spring consumption.

If the fall flows are uncertain and little fall surplus honey has been gathered the bees will have to be fed a sugar syrup for their use during the winter. The minimum amount of stores (either honey or sugar syrup) necessary for the winter consumption of the colony is fifty pounds. The beginner will find it convenient to take a pair of scales, similar to ice scales and weigh his colony of bees. A complete hive with its bees should weigh about forty pounds and any weight over that may be considered as stores. For example: if a colony weight is seventy-five pounds then it has thirty-five pounds of honey. To gain the additional fifteen pounds necessary for minimum stores sugar syrup should be fed to the colony. The best feed for winter stores is made by mixing two parts sugar to one part warm water. A gallon of this mixture increases the weight of the colony about seven pounds so a colony weighing seventy-five pounds would need two gallons of sugar syrup to bring it up to the minimum stores for successful wintering.

Common practice is to leave a second hive body on each colony, one with sealed stores of honey and pollen being desirable.

UNITING WEAK COLONIES

Occasionally winter arrives with some of the colonies in a weakened condition. This may have been caused by a failing queen, or by a bad season with a minimum of honey for the bees to gather. If such a condition has been discovered early enough so that new queens could be supplied, or ample feed provided, or both, it may be possible to pull such colonies out of their doldrums and make them fit for wintering. Even with the best of protection, these under par colonies will have difficulty surviving.

Generally it is better to throw all the strength of two such colonies into one hive by uniting than to try to winter the two misfits. In a honeyflow uniting gives no difficulty. Nectar gathering bees unite very nicely with a minimum of fighting. It is a different matter in the short days of fall. In such a case the two colonies may be united by the newspaper plan.

Find the poorer of the two queens in the two hives to be united, and kill her. Or if uncertain, do not bother. The bees will see to a proper choice, or the queens will fight it out for themselves. Remove the cover and inner cover of one of the hives, exposing the tops of the frames. Place immediately over these a single sheet of newspaper. Over this place the second colony with its bottom board removed, and the job is done. In the course of a few hours, the bees of both colonies will have gnawed through the newspaper, getting acquainted a bit at a time, and amicably. After they have been completely united, it may be best to remove one of the supers or hive bodies. This can readily be done by smoking the bees down, or shaking them into the brood chamber which may consist of only one hive body; preferably two for winter.

With a good queen, plenty of young bees, and careful attention to the amount of available stores the beginner should have little or no trouble in bringing his colonies through the winter.

As plant life becomes dormant, brood rearing and the field activities of the colony decline until they finally cease entirely. The population of the colony decreases rapidly in the fall until only bees from four to six weeks old are left in the hive. Since they have done little or no work these young bees are capable of living through the winter.

At a temperature of around 57 degrees F. the bees start to form their winter cluster. When the temperature has dropped as low as 43 degrees F. all of the bees have joined the winter cluster. With the formation of the cluster one of the first indications of approaching winter is the action of the colony in relation to its male population- the drones. The worker bees chase the drones out of the hive where they soon die from starvation and cold. Since the drones have no useful task in a normal functioning colony they are a useless burden most of the time. The worker bees tolerate the drones during the spring and summer when there is a chance that they may be needed to mate with a queen, but the first hint of cold weather is the signal for a general forced exodus of the drones.

There has been much controversy in this country on the question of whether to pack bees or not to pack them. For many years the standard practice was to pack the hives in a heavy coating of straw, leaves, or some other insulating material. The material used to insulate was usually held in place by a tar paper pack which went around, under, and over the colony. The theory was that this packing would so insulate the colony that the natural colony warmth would be retained within the hive and allow the bees to move freely to their food reserve wherever it was stored in the hive. Some even insulated too heavily with the result that during warm days in the winter the colony was so warm that it was forced to move outside the hive

During the last few years there has been a general trend away from the heavy packing and insulation of the colony during winter. Lighter packing

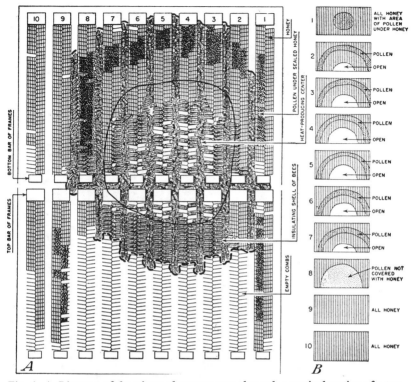

Fig. 4. *A, Diagram of the winter cluster as seen through a vertical section of a two-story hive cut across the middle of the cluster. B, Face view of frames of upper hive body. The numbers indicate the position of the frames. Note how the bees concentrate between combs and in open cells to form a compact insulating shell around a much less compact heat-producing center. The band of pollen covered with honey indicates an accumulation of reserve pollen before the honeyflow.*

was tried and when that proved successful, colonies were wintered with no packing. The unpacked colony wintered as well as the packed colonies had done and today the majority of the commercial beekeepers do not pack their colonies.

Whether the colony is packed or not packed, two of the requisites for successful wintering are a reduced bottom entrance and a small top entrance. The entrance closer supplied with the original hive should be placed in the bottom entrance of the colony leaving only the smallest opening for the bees to fly from. It should not be necessary to close the entrance down so far until cold weather has set in for the winter. In addition to the reduced bottom entrance the bees should be supplied with a top entrance. A one-inch auger hole is bored just below the hand hold in the front of the super which is left on the colony. This top entrance allows the bees to fly freely from the top of the cluster during days when the temperature rises to 45 degrees or higher and also allows some circulation in the hive to help dissipate excess moisture. If the bottom entrance should become choked with dead bees or other material the top entrance will then act as a safety exit from the hive.

Fig. 5. A close-up view of the top entrance in early autumn. On warm days in the winter the bees will fly from this entrance before using the bottom one.

The beginner will have to use his own judgment about wintering his bees. If it is felt that the climate is too cold and severe for successful wintering with unpacked colonies then by all means take the time and trouble to place some kind of light packing around the hives. But remember - packed or unpacked always use the reduced bottom entrance and the auger hole top entrance.

Beekeepers in many localities are bothered during the winter by field mice. The mice enter through the lower entrance and remain in the hive during the winter, usually chewing up the combs during their stay in the hive. If mice abound in the locality, it may be well to use some sort of a guard at the entrance to the hive. Most beekeeping supply manufacturers carry a metal entrance reducer for this purpose.

Although some beekeepers ignore wind protection, there are many who prefer to have some protection from prevailing winter winds. This is accomplished by several means: placing the hives on the southern slope of a hill, in

Fig. 6. A. Feeding 2:1 sugar syrup to dispense drugs and provide adequate stores.

B. Insulite board over inner cover to serve as moisture-releaser. Note upper entrance and reduced lower entrance.

C, D. Installing commercial cardboard winter pack.

E. Colony ready for winter minus upper entrance.

F. Hole bored in cardboard pack to correspond to upper entrance in B.

Fig. 7. A wooden fence used as a windbreak for wintering bees. Such a fence will also serve to direct the line of flight of bees upward, rather than directly across neighboring properties.

a grove, or putting a snow fence or some other windbreak between the hives and the direction of the prevailing winds. Natural windbreaks are best since they entail no extra management or work on the part of the beekeeper.

If no windbreak is available, moderate packing is likely desirable. Remember, that with a packing case, a small entrance should always be provided, so the bees may fly on moderate days during the winter thus being able to remove the dead, and to void themselves of accumulated excrement, since bees never void themselves within the hive. Generally, no packing is considered necessary in the southern tier of states.

The conditions necessary for successful wintering may be put down in brief form, as follows:

(1) A good queen early in the fall.
(2) Plenty of young bees.
(3) A minimum of fifty pounds of stores—honey, sugar syrup, or both.
(4) Reduced lower entrance and a 1-inch auger hole in super.
(5) Protection from prevailing winter winds.
(6) Exposure to winter sunshine.

CHAPTER XII

SPRING MANAGEMENT OF THE OVERWINTERED COLONY

EARLY SPRING MANAGEMENT

I F THE bees were packed for the winter, the first early spring job (late March or early April*) is to remove the packing and dispose of it. Next, whether the bees were packed or not, tilt the hive back, remove the bottom board and scrape it clean of debris or dead bees. Replace the hive on the bottom board and be sure to replace the entrance reducer. Bees will rob readily in the spring and they need the reduced entrance to help them guard the hive.

During the first spring examination of the overwintered colony check for evidence of a laying queen, supply of honey and pollen stores. If the colony has a laying queen the brood rearing will start comparatively early-usually sometime in February or March. If an examination shows the colony to be strong in worker population, but queenless, order a new queen at once so that she will arrive in time to place in the colony during the first of the fruit blossoming period. As both pollen and honey are necessary for brood rearing, adequate supplies of both must be available if the colony is to build up in strength for the main honeyflow. If wintering instructions were fol-

Fig. 1. Brood development in March in a strong colony that has ample stores of both pollen and honey.

---

* All references to seasonal dates refer to central Illinois. Spring moves northward at a rate of approximately ten miles per day. If you use this figure you will be able to change the dates mentioned in this text to conform with your particular locality.

lowed carefully there should be plenty of honey for the colony at the time of this first examination. It is easy to tell when a colony is short of food. A safe rule is to make sure that there is some sealed honey in the comb at all times. When there is no sealed honey to be seen, the bees are approaching starvation. If the colony is short of honey it should be fed sugar syrup. (Sugar syrup is prepared by mixing one part of sugar to one part of hot water, allowing the mixture to cool before using.) With no honeyflow on and brood rearing increasing, a colony will consume approximately ten pounds of sugar syrup during a one-week period. Continue feeding sugar syrup until the natural supply of available nectar is sufficient for colony needs.

Considering that pollen is essential for brood rearing, the early spring period, when natural pollen either is not available or inclement weather prevents the bees from gathering it? is a critical time in the build-up of the colony. Check each comb carefully for pollen reserves. If the supply of pollen is low, feed a pollen substitute. Pollen substitute is made by mixing one part animal type brewer's yeast with two parts of expeller-processed soybean flour by weight. One-to-one sugar syrup is then added to the dry mixture until it is of a paste-like consistency, just short of being runny. Spread about three pounds of this mixture over the tops of the frames directly above where the bees are clustered, and cover with wax paper to keep the pollen substitute from drying and becoming hard. Continue to feed freshly prepared pollen substitute in this manner until natural pollen becomes available.

Fig. 2. Pollen cake being fed to a colony in early March. (Photo Univ. of Wisc.)

It is recommended that Terramycin* (*TM Pfizer) be added to each pollen substitute cake and each feeding of syrup during the spring period. The use of this drug serves as a very effective preventive for American foulbrood.

MIDSPRING MANAGEMENT

The period of fruit bloom—starting with the apricot and pear, and ending with the last of the apple blossoms—bringing with it an abundance of natural pollen and nectar, is a period of great increase in colony strength. Stronger colonies may be given a full entrance at this time, although it would still be advisable to keep the entrances of weak colonies somewhat reduced. It is during this period that the results of a queen's egg laying should be full combs of solid brood. You should, therefore, watch closely for poor queens—distinguished by their spotty or small amount of egg laying. Such a queen should be replaced at once so that the new queen will have a chance to build up the colony during fruit bloom.

If one of the colonies seems to have a good queen but is weak in worker population, it may be helped by exchanging an empty comb from the weak colony with a good comb of emerging brood and bees from a strong colony that can spare the loss of a few bees. When doing this, check the

Fig. 3. A large well-developed queen. Notice that the queen has been marked for easy identification. A nominal charge is made by most queen breeders for this service. This queen is laying a good pattern of solid worker brood.

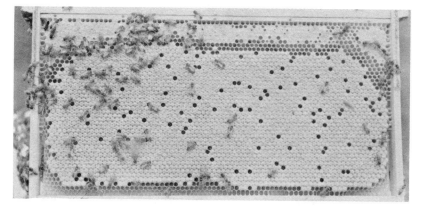

Fig. 4. An excellent comb of sealed brood. Very evidently the work of a superior queen. A comb of brood like this will very quickly bolster the strength of a weak colony or build up a package colony.

comb closely to make sure that you are not taking the queen from the colony. Combs of brood added in this manner should be placed on one side of the brood nest so there will be little danger of the strange bees killing the queen.

The worker bees of a colony which leave the hive to gather nectar, pollen, or water are called field bees. They are accustomed to the position of their own hive in an apiary and should the hive be moved to a different location they would return to the old location rather than to the hive itself. This habit of the field bees gives still another method of building up a weak colony - switching. In this method a weak colony is exchanged in apiary location with one which is strong. Since the field bees will return to their old location, the weak colony will gain considerable strength in numbers. This should be done when bees are gathering nectar.

As the fruit bloom period progresses, many colonies will have brood in the super which was left on for winter stores, while the hive body of combs below will be only partially used and in some cases may be entirely deserted. If this should be the case, reverse the position of the body and the super-placing the super below the body. This will cause the queen and bees to move up into the partially deserted hive body where there is plenty of room for the queen to lay her eggs. If there is honey left in the super, the worker bees will move it up into the body and in so doing will stimulate the colony to faster growth.

An occasional colony will be found in which the queen confines her egg laying efforts to the center combs even though there may be side combs which are completely empty. This is especially true if there are frames of

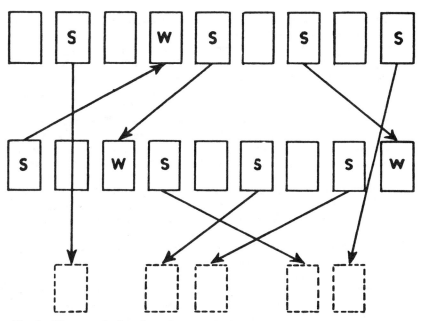

Fig. 5. *Diagram of relocation. Strong colonies are indicated by the letter S; weak colonies by the letter W. Arrow indicates possible relocation.*

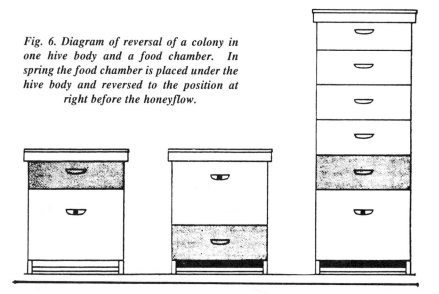

Fig. 6. *Diagram of reversal of a colony in one hive body and a food chamber. In spring the food chamber is placed under the hive body and reversed to the position at right before the honeyflow.*

Fig. 7. left to right; top to bottom. Diagram of expanding the brood nest. With four fairly well-filled brood combs flanked by stores in fruit bloom, one empty comb is moved in; a week later, another is moved in, and about 10 days later two empty combs are moved in as diagramed.

honey or pollen blocking the expansion of the brood nest and effectively preventing the queen from laying in all of the combs. In such a case the empty side combs should be moved up to the brood nest and the combs of honey or pollen moved toward the sides of the hive. Do not, however, separate the frames of the brood area. Remember that word of caution— always put frames of combs up to and not into the brood nest.

LATE SPRING MANAGEMENT
From the end of the fruit bloom period until the beginning of the major honeyflow there is usually a period of dearth, often between the blossoming of dandelion and clover, when there is little or no nectar gathered by the bees. Many colonies which did not need additional feed during the early spring or midspring, may need syrup during this period. Some colonies may need only one feeding of syrup to carry them through, while others may

require as many as three ten-pound feedings of syrup. Here again it is advisable to add 1/2 tsp. Terramycin TM 25 to each ten pounds of syrup given to a colony.

This period is truly called the danger period since many cases of colony starvation and swarming occur at this time. Starvation is easy to prevent and the wise beekeeper will make sure that his colonies have plenty of natural stores or sugar syrup during this period.

Swarming presents a number of different problems, all of which may be overcome by the beekeeper. Chapter VIII, "Swarming," presents a discussion of swarming problems and their solutions.

Although some beekeepers talk proudly about having all of their equipment full of bees, the good beekeeper will have some extra equipment which is available at all times. It would indeed be poor management to allow a colony to swarm simply because there was not an extra hive and a few pounds of comb foundation necessary to divide a colony. By the time the equipment could be ordered and arrive at its destination the swarm would have long since departed.

RELOCATION

An easy and very effective measure of swarm control is the practice of relocating or switching colonies. As described previously, the weak colonies exchange places with the strong. Not only does this cut down the field force of the crowded strong colony, and hence help to eliminate swarming, but it also helps to build up weak colonies that would otherwise not be able to make much of a honey crop. If no weak colony is available, the strong colony is simply relocated in a new spot so that the field bees will drift away from it.

DIVISION

As a last resort, the colony attempting to swarm may be divided into two parts and the queenless part given a new queen. When making the division, it is advisable to give a majority of the sealed brood to the new queen, leaving the old queen and part of the brood at the old location. The new divide can be placed three or four feet from the parent hive and the new queen introduced. Most of the field bees will return to the old location, leaving only the new queen, the young worker bees, and the brood in the new hive. They may be run as two separate colonies during the honeyflow or may be united as one colony shortly after the start of the main honeyflow. In the latter case, frequently one queen will disappear shortly after uniting or, both queens will continue to lay until the end of the honeyflow when one of them will disappear.

WAYS TO MAKE SPRING INCREASE—PACKAGE BEES

The established beekeeper is in a far better position to make increase with package bees than one who is just beginning with his first few colonies because he can use his established colonies to help his package colonies. Packages should arrive as early as possible, preferably during the first half of April, earlier or later according to location. Fruit bloom will just be starting at this time. Three weeks after the packages are hived, during the period of their lowest ebb, they should be bolstered if possible by the addition of a comb of sealed brood and the adhering bees. If old colonies can spare it, the new packages should have a second comb of brood and bees added some time during the fourth or fifth week of their growth. Packages taken care of in the above manner, and supplied with an ample amount of syrup and pollen, will gather a fair honey crop the first year.

INCREASE BY DIVISION

As previously described, strong colonies can be divided and the queenless half given a new queen. It is also possible to take combs of sealed brood with the adhering bees from strong colonies, replacing them with empty drawn combs or frames containing full sheets of comb foundation. These combs of sealed brood need not come from the same colony but may be taken from a number of colonies. Three frames of sealed brood and bees, given a new queen the first half of April, will grow into a good colony for the honeyflow. Later, divides may be made by taking four or five frames of brood and bees and introducing a new queen the first of May. It is best to place the brood and bees in the hive in a new location and to wait twenty-four hours before introducing the new queen.

The serious beginner will now want to expand his number of colonies from two or three on up to ten or more. For the man just starting in with bees, two or three colonies is enough to occupy his spare time. There are many things to be learned by one just starting with bees. You have finished an entire season by now and are more adept in frame and colony manipulation. Tasks which formerly took an hour to perform are now simple ten-minute jobs which come to you naturally. Brood, queens, workers, drones, all are stamped in your memory and a quick glance tells you what you want to know. The results of a good queen are quickly discernible and no longer must you ponder over a textbook to discover whether a queen is good or poor. All of this knowledge you have accumulated and all of your increased adeptness in manual manipulation mean that you are now able to care for at least ten colonies in the same length of time in which you formerly cared for only two.

CHAPTER XIII

DISEASES AND ENEMIES

FOULBROOD

O F ALL the diseases of bees the most dreaded is foulbrood. Foulbrood attacks the larvae, which die in the cells. It is infectious and must be treated with promptness and care. There are three kinds of foulbrood-American, European, and para. Of these, American foulbrood is the worst scourge and requires the most radical methods of treatment, since the bacteria *(Bacillus larvae)* are transmitted easily from one colony to another through the honey or through the exchange of combs, or through careless handling by the beekeeper. European foulbrood is a disease of the brood that confuses beekeeper and researcher alike. The serious effects include crawling bees and lowered colony strength. Parafoulbrood, while possibly more easily transmitted than either of the other two, occurs rarely, being limited at present to southeastern United States.

American foulbrood attacks the brood in its larval stage. The larva at first becomes light brown, changing to dark brown and finally to dark coffee-color, sinking in a mass to the bottom of the cell with a tongue-like attachment to the upper side of the cell. As it dries it becomes sticky like glue, with the offensive odor of the gluepot. On insertion of a small stick into the mass it will often stretch out one or two inches in a ropy manner.

The scales, when dried, adhere to the cells so closely that the cell walls must be torn down to get at the dried mass. When the bees try to clear out the disease they may, on account of its tenacity, spread it instead; although experiments have shown that some bees are more resistant to the disease than others and that some may clean it up altogether. Perhaps careful selection through many generations may develop a strain of bees capable of resisting the disease. A good start has already been made along these lines.

Very often the bees seal the cells of the larvae either before or after the larvae become diseased. As the disease develops, the larvae die; the cell cappings become punctured and sunken. An advanced case of the disease may show a comb with most of the brood affected, ranging all the way from open larvae just freshly diseased through sticky masses at the bottom of the cells, cells with punctured cappings, and cells with dried scales.

This disease is transmitted both through the combs and through the honey, probably chief through the robbing out of weak or dead colonies or through the careless exposure of such combs and honey during a nectar dearth. Undoubtedly a common means of spreading the disease is by the use of equipment in which diseased colonies have died. Extreme care should be taken to scorch with a flame all such equipment about which there is any doubt.

Fig. 1. American foulbrood in all stages, ranging from open discolored larva to cells with dried scales. The open cells contain sticky, decaying larva.

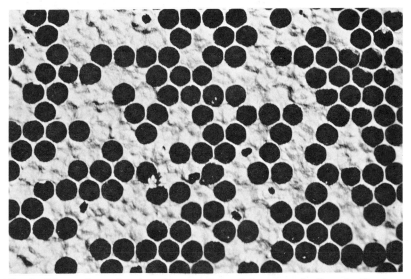

Fig. 2. The perforated and sunken cappings of American foulbrood. The bees have removed larva from the open cells, but many of them still contain dried scales.

Fig. 3. A cell, believed to have been killed by American foulbrood is opened and the contents probed with a twig or matchstick. The strong smell of decay and the ropiness with which the decayed larva will string out of the cell will almost certainly indicate AFB.

Many states require the burning of all material infected with American foulbrood. Also many states require that colonies moved into the state be accompanied by a certification that such colonies are free of disease.

Diagnosis of the disease is a service available to the industry by the U.S. Department of Agriculture. The sample of comb should be 4 or 5 inches square and contain as much of the dead brood as possible. Mail it to the Microbiologist, Bioenvironmental Laboratory, Beltsville, Maryland 20705. There is no charge for this service.

The use of Terramycin® for the prevention of American foulbrood has been successful in many cases. Use 1/2 teaspoonful of TM 25 to 3/4 teaspoonful of powdered sugar. Dust a teaspoonful of the mixture over the brood combs 2 to 3 times in early spring, with intervals of 7 days to 10 days between each application.

European foulbrood became at one time a veritable scourge among the beekeepers of the United States. Fortunately the method of abatement of the disease is simpler than those used in treating colonies infected with American foulbrood. The disease is caused by Streptococcus pluton.

The larva soon after hatching from the egg assumes a yellowish color which later changes to grayish or blackish. Because of its appearance at this stage the disease was at one time called black brood. The larva usually dies before the cell is sealed, and it never adheres as closely to the cell wall as the dried scale of the larva infected with American foulbrood does, so that the bees are able to carry out the dead larvae, which they rarely do in the case of the other disease. But if the disease is not treated, the colony weakens and the moths invade it, eating only the wax, not the dead brood.

There is no ropy brood; or if there is any ropiness it is of small extent. There is usually no marked foul odor except a slight, sour odor of decaying flesh, a real carrion odor caused by the existence of other bacilli of putrefaction, such as Bacillus alvei, which put in their work after Streptococcus pluton has killed the larvae.

As European foulbrood is readily cured by strong colonies of Italian bees it is not necessary to destroy the combs. The queen is removed and, if she is not of the best, she is killed. Within ten to twenty-one days the bees will have cleaned out the dead larvae and then may be given a healthy young Italian queen. If the colony is strong the disease will disappear. Sometimes caging the queen in the hive from ten to twenty days is sufficient, but if the colony is weak it may be best to unite it with another. We prefer to destroy the old queen, and to introduce a new one, as cases have been known of the old queen carrying the disease to another colony if introduced to it.

Drug Treatment - The drug used for the control of European foulbrood is Terramycin,® TM 25 or TM 10 (use 2-1/2 times the following amounts if using TM 10).

It is advisable to use Terramycin® as a dust rather than in syrup because, in liquid, Terramycin® loses its potency quickly. Use 1/2 teaspoonful of TM 25 to 3/4 teaspoonful of powdered sugar. Dust a teaspoonful of the mixture over the brood combs three times between the beginning of brood rearing in spring and the beginning of the honeyflow.

Another effective way to treat colonies for the control of American and European foulbroods is with the use of extender patties. The advantage of using this method is that it has also been shown that the use of extender patties also aids in the control of trachael mites. The recipe for these is simple:

1-6.4 oz. pkg. terramycin soluble powder
4.6 lbs. - vegetable shortening (Crisco, etc.)
9.12 lbs. table sugar or powdered sugar

The above should be mixed thoroughly. The recipe will yield 14-1lb. patties. Other size batches may be made proportionately. Sandwich the

Fig. 4. Closeup view of European foulbrood damage. Some of the larva appear almost normal, others are infected and discolored, and some are dead masses in the bottom of the cell.

patty in wax paper and place on the top bars above the brood nest. Use 1 pattie per colony when there is no honey flow.

Parafoulbrood occurs so seldom that only brief mention of it is necessary. It has some of the characteristics of both American and European foulbrood. The treatment recommended for it is the same as for American foulbrood.

It is an offense punishable by fine to allow foulbrood to exist without treating it. There is no doubt that it can be destroyed by the methods suggested. Any beekeeper who is afraid to treat colonies without help should write to the state inspector of apiaries.

NOSEMA DISEASE

This is a disease of adult honey bees caused by Nosema apis, a small protozoan, which multiplies in the digestive tracts of adult bees and shortens their life as much as 50 per cent. Nosema may cause the loss of many adult bees and reduce the population of colonies without the beekeeper knowing it is present. In bad cases, especially during the first few days of the honeyflow, many bees with Nosema may be found about the bee yard. The wings are spread, legs unsteady, and intestines swollen and discolored. The stomach, normally dark, is yellow or white.

Nosema disease has caused the loss of many queens through supersedure of infected queens especially in package bee colonies. In package bees, when infection with Nosema is high, it will seriously affect the death rate of the bees even if the queen escapes infection.

Naturally, strong colonies with plenty of stores of honey and pollen and with young vigorous queens are in best condition to combat the ill effects of Nosema.

Since Nosema germs appear most prevalent in stagnant brackish water, nearby sources of fresh water for the bees' use are highly desirable.

Drug control —Fumidil-B,® a new drug developed by Abbott Laboratories, is very effective in the control of Nosema. Dosages are so well defined on the packages and containers that it is not necessary to give them here. In with each package of Fumidil® is a small booklet with details about the drug.

OTHER DISEASES

Sacbrood is similar to European foulbrood but milder. The dead brood remains in a sac-like form which can be removed, intact, with a toothpick. It is readily removed by the bees. Usually the disease disappears of itself during a honeyflow. Requeening with good vigorous stock is a help.

A disease which has rendered much havoc in the British Isles is Isle-of-Wight disease which affects the trachea of the bees. It is of no importance in North America.

In order to reduce to a minimum, the possibility of the importation of bee diseases from other countries, regulations of the U. S. Department of Agriculture do not allow the importation of any bees or queens into the United States.

Bee diarrhea in the latter part of winter and early spring is a malady that effects some apiaries. Over the hives and combs, the bees discharge water excrements of a dark appearance and an offensive odor. The cause is either fermented honey, improper food, long confinement, or too warm and poorly ventilated quarters.

Fruit juices harvested during a dearth of honey is the food that most frequently causes diarrhea. When these juices, not sweet enough to keep from fermenting, are stored in the combs like honey they should be extracted and replaced with good honey or syrup. Honeydew, from plant lice, is also a cause of diarrhea when cold weather confines the bees to the hive for a long time.

Usually diarrhea disappears with the first flights of the bees. But in a protracted winter it is difficult to cure. It is much more easily avoided by pure food given at the opening of winter than stopped after it has once begun. When syrup is fed none but the best granulated sugar should be used. The feeding of 2 gallons of syrup (2:1) is the best substitute for honey stores.

PARASITIC BEE MITES

Honey-bee Tracheal Mite - Acarapis woodi

The honey-bee tracheal mite, *Acarapis woodi*, affects only adult honey bees. The parasite was first described in 1921 in bees in Great Britain (Rennie 1921).

The honey-bee tracheal mite can be positively diagnosed solely on habitat as it lives exclusively in the breathing tubes (prothoracic tracheae) of the honey bee.

The female tracheal mite is very small and can only be viewed through a microscope. The body is oval, widest between the second and third pair of legs, and is whitish or pearly white with shiny, smooth cuticle; a few long hairs are present on the body and legs. It has elongate beak-like mouthparts with blade-like styles for feeding.

When over 80 percent of the bees in a colony become parasitized by the tracheal mite, honey production may be reduced and the likelihood of winter survival decreases with a corresponding increase in infestation (Bailey, 1961, Furgala et al, 1989). Individual bees are believed to die because of the disruption to respiration due to the mites clogging the tracheae, the damage caused by the mites to the integrity of the tracheae, microorganisms entering the hemolymph (blood) through the damaged tracheae, and from the loss of hemolymph.

The mites are transmitted bee to bee within a colony by queens, drones and workers. In addition, the interstate transportation to package bees and queens, as well as established colonies, has resulted in the dissemination of the mite throughout much of the United States.

The population of the tracheal mite in a colony may vary seasonally. During the period of maximum bee population, the percentage of bees with mites is reduced. The likelihood of detecting tracheal mites is highest in the fall and winter. No one symptom characterizes this disease; an affected bee could have disjointed wings and be unable to fly or have a distended abdomen, or both. Absence of these symptoms does not necessarily imply freedom from mites. Positive diagnosis can only be made by microscopic examination of the tracheae; since this mite is only found in the tracheae, this is an important diagnostic feature.

Acarapis woodi, the honey-bee trachael mite

There are several techniques for dissecting and examining bees for the tracheal mite. Since all of these techniques require specific training, equipment and chemicals, please contact your State Apiarist for a complete and accurate diagnosis.

There are control methods which are available at this time for the control of the Honey-bee Tracheal Mite. **Please call your nearest Dadant and Sons office for availability and information on the proper use of the chemical and nonchemical controls of the Tracheal Mite.**

The variations in the effect of the tracheal mite on individual colonies suggests that breeding bees that are tolerant or resistant to the mite is feasible (Gary and Page,1988). The development of such bee stocks that still have all the desirable traits plus resistance to mites is the method of choice over the use of chemicals. However, this may take several years. Meanwhile, beekeepers can minimize the impact of tracheal mites by intensive management practices to maintain populous colonies.

Varroa jacobsoni

The parasitic bee mite, *Varroa jacobsoni*, is one of the most serious pests of the honey bee. The mite has become established on every continent except Australia and will continue to spread because of the commercial transport of bees and queens, the migratory activities of beekeepers; swarms that fly long distances, or are carried by ships or aircraft, and drifting bees.

The adult female Varroa is large, pale to reddish brown in color and can easily be seen with the unaided eye. Male mites are considerably smaller and are pale to lightly tanned. Adult bees serve as intermediate hosts when little or no brood is available and as a means of transport. The females attach to the adult bee between the abdominal segments or between body regions (head, thorax, abdomen), making them difficult to detect. These are also places from which they can easily feed on the bee's hemolymph (blood). The adult bee suffers not only the loss of blood, but may be subjected to microbial invasion, leading to a reduced life expectancy (DeJong and DeJong, 1983).

The most severe parasitism occurs on the older larvae and pupae, drone brood being preferred to worker brood (Ritter and Ruttner, 1980). The degree of damage depends on the number of mites parasitizing each bee larva. One or two mites will cause a decrease in vitally of the emerging bee. Higher numbers of Varroa per cell result in malformations like shortened abdomens, mis-shaped wings, deformed legs or even death of the pupae.

Varroa can be found on the adult bee, on the brood, and in hive debris. In looking for Varroa, remember that the number and location of mites in a colony vary according to time of year,the number being lowest in the spring, increasing during the summer and highest in the fall. During the spring and summer most mites are found on the brood (especially drone brood). In the late fall and winter most mites are attached to adult worker bees.

Ether spray can be used to dislodge Varroa from adult bees.This technique is a rapid and efficient field detection method and avoids the handling, shipping and time-consuming procedures associated with other detection methods. The bees (50 - 100) are collected in a jar and anesthetized with ether delivered from an aerosol can (this aerosol formulation is easily obtained in auto parts stores where the product is sold as an aid to start engines.) A one to two second burst of material is adequate. The bees are then rotated in the jar for about 10 seconds, hence the name ether-roll technique. The mites, the majority of which will have dislodged from their hosts, stick to the inside wall of the jar. To complete the process, the bee sample is deposited on a white surface and spread around. The bees should be examined immediately after the ether roll since the mites tend to stick to the bees if left in the jar more than a few minutes.

To look for mites on brood, pupae (preferably drone) are examined. Varroa can be easily recognized against the white surface of worker or drone pupae. Pupae can be removed from the cells with a cappings scratcher, forceps or hive tool. However, a minimum of 100 pupae should be examined per colony. A quick and easy method of obtaining pupae is to insert a cappings scratcher at an angle through the cappings and lift the pupae and cappings upward. Another method is to slice off the caps of brood with a long-bladed knife. The comb (frame) is then sharply jarred on a hard, flat surface such as a hive top; this dislodges the brood on which the mites are easily observed. Also be sure to examine the bottom of the cells which once held the pupae.

Hive debris (wax particles, pollen, dead bees and brood, mites, etc.) normally fall to the hive floor and are removed by housecleaning bees during warm weather. This material can be collected and examined for the presence of Varroa. A magnifying glass or dissecting microscope can be helpful in locating the mites in the debris. The collection of hive debris can be facilitated by placing white construction paper or cardboard on the floor. Stapling the paper under a wood (1/4inch) and wire (8 to 12 mesh) frame protects both the paper and debris from the bees. The paper is examined for mites that can easily be seen against the white background.

Infested honey-bee colonies will die sooner or later if the beekeeper does not effectively control Varroa. There are chemical control methods which are available at this time for the control of the Varroa mite. Please call your nearest Dadant and Sons office for availability and information on the proper use of these chemicals for control of the Varroa mite.

ENEMIES

The enemies of bees are not numerous. A few birds, among which we shall name the kingbird, eat bees. But their damages are so insignificant that they are hardly worthy of mention.

Ants sometimes make their nest over the frames to take advantage of the warmth of the bees. They may be driven away by placing salt or powdered sulphur where they congregate. The bee louse, Braula coeca, and the death's head moth, which enter the hive to feed on the honey, exist in Europe and the Argentine but are practically unknown here.

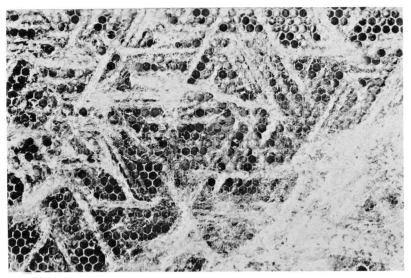

Fig. 5. A comb completely destroyed by the criss-crossed webs of the wax moth.

The most active enemy of bees is the bee moth, Galleria mellonella, which lays its eggs in neglected combs, especially in old combs. The larvae hatch and devour everything in their reach, making webs or galleries through the combs. Colonies that have more combs than they can cover or queenless colonies, especially in the fall when the moths have already reared two broods and are therefore numerous, are often rendered worthless by the ravages of the larvae of the bee moth. Two different kinds of moths are known but the larger, Galleria mellonella, is the principal depredator. Luckily, the bee moth cannot stand a temperature below zero. It is only when sustained in some corner of a populous colony or in combs in a warm room that the moths can remain alive to reproduce from one year to another.

There is but one remedy - if colonies are kept strong they will destroy the moth. Combs should never be kept in exposed places. When moths are discovered in stored combs, they may be destroyed with paradichlorobenzene.

Fumes of paradichlorobenzene are heavier than air and descend so the stack of supers should be made tight. On top of the stack, put a handful of paradithlorobenzene on a paper. See that all is covered tightly with enough space around the chemicals so the fumes may readily diffuse and descend through the combs to be treated. The combs should be inspected monthly for signs of infestation, especially if the temperature rises above 60º F. Before using the combs again, they should be aired well to remove all chemical fumes.

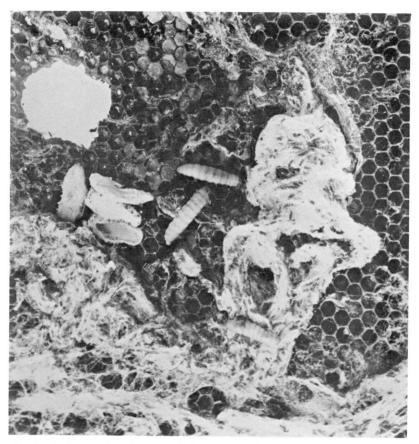

Fig. 6.   A closeup view of wax moth damage, showing both the webbed tunnels and the wax moth itself in the larval stage.

CHAPTER XIV

QUEEN MANAGEMENT

REQUEENING

THE BEGINNING beekeeper will do well to buy bees which have been carefully bred. The reason for this is simple. It is possible to breed for certain characteristics desirable in bees which are to be used in honey production for commercial purposes. Bees should possess hardiness to withstand the rigors of the climate to which they are exposed, also hardiness to withstand the possible ravages of bee diseases and enemies. Bees should be gentle in order that they may be handled with maximum speed and efficiency. They should remain quiet on the comb and give little trouble when being examined. For an apiarist located in or near a city the gentleness of his bees is most important. Cross bees that sting not only disturb the beekeeper, but they bring angry neighbors and possibly outside legal action. Bees should be industrious and should breed to strong colonies for the honeyflow. They should be out early and late for nectar pollen and water. Their industriousness will add to the possibility of their gathering from every floral source. Not all strains possess these desirable traits. Indeed, some of the characteristics have not yet been possible to fix; but scientists are making progress in this field.

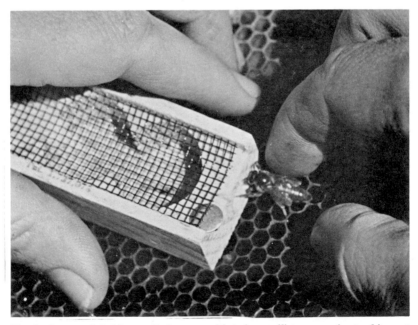

Fig. 1. A young hybrid queen being coaxed into her mailing cage prior to shipment to a beekeeper where she will be used in requeening a colony.

Recently, hybrid bees that are gentle, better honey harvesters and have other desirable characteristics have been developed by Dadant & Sons, Inc. at Hamilton, Illinois. The first hybrid was known as the Starline hybrid and was well received for its exceptional brood-rearing ability and resultant performance in increasing yields of honey. The next hybrid was called the Midnite hybrid, a darker bee with Caucasian background, and excelled in gentleness and also was a good gatherer of honey.

The breeding stock for these hybrid bees is made possible through artificial insemination, a technique developed by Dr. Lloyd R. Watson and perfected by government scientists. This enables selection, through repeated and long testing of performance, of breeding stock that provides bees with desirable characteristics. The experimental work is being continued to improve the hybrids each year, and time will see honey bees bred for special purposes e.g. the pollination of specific crops.

Fig. 2. This vigorous young hybrid queen is the product of careful selection for gentleness, egg-laying ability, and her off-spring will also be excellent honey-gatherers. (Photo, E. R. Jaycox)

Because the future of a colony of bees depends solely on the queen it is imperative for best results that bees be bought with queens of good stock or that proper requeening be done.

Requeening means changing the queen of a colony, either to secure bet-

Fig. 3. The best check on a queen is the pattern of the brood. The scattered brood and the bullet-shaped cells give her away as a drone-layer. She should be replaced with a new queen.

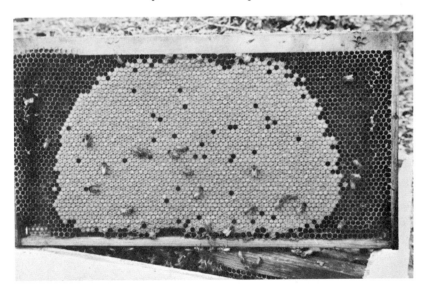

Fig. 4. This queen is laying in a good solid concentric pattern. The flat, even cappings indicate that the cells contain worker larva.

ter stock, or to replace a queen which has passed her peak in volume of egg laying and consequently cannot furnish the eggs necessary to the development of a normal colony. In nature the bees replace failing queens themselves. In such cases, however, it may be attended more or less by disorganization of the colony and perhaps considerable swarming. Many beekeepers, therefore, do not wait for colonies to replace their queens by natural means.

Colonies are requeened usually during a light honeyflow, since queens are better accepted then. If a colony is found with a failing queen in the spring, the earlier she can be replaced the better. Most requeening is done just at the close of the major summer flow. This gives the new queen time to be accepted and to rear a rousing colony of her own bees before the winter season begins. Requeening for improvement of stock may be done at any time that queens may be obtained, but preferably when some honey is coming in.

Lately some apiarists have practiced requeening during the broodless period of late fall, claiming better acceptance at that time. This method should succeed with strong colonies, as there are ample bees to carry through winter. Earlier requeening of a colony which is losing strength during late summer and fall, might be more satisfactory, since it would allow the new queen to replenish her worker bee force before winter sets in.

How often should requeening be done to replace possible failing queens? The beekeeper will have to be the judge. Much depends on how much strain has been put on the laying activities of the queen. In a climate of short season and short flows colony maximum strength has to be maintained only for a relatively short period. On the other hand in a prolonged season, particularly one of heavy honey production and heavy egg laying, the queen wears out much faster. Some commercial producers requeen annually. Probably most producers requeen every other year. Some beekeepers requeen only as needed believing bees are better judges of their queens than they are and/or that the queen they introduce may not be a better one than they removed.

To introduce a queen successfully it is necessary to find the queen to be replaced and dispose of her. When the hive is smoked and opened, some queens run on the combs and even onto the walls of the hive making it extremely difficult to find her. Therefore, it is best to proceed cautiously and without jarring the hive.

In the middle of the day, when the old bees are at work, the hive may be opened. It is well to begin with the center frame, examining both sides and, if the queen cannot be found, to proceed with the adjacent frames until she is found. If she seems to be completely hidden and it is impossible to find her, the hive should be closed for an hour or two, until the bees have become quiet once more. At that time the operation may be repeated.

When the queen is found, she should be destroyed or such disposition made of her as may be desired. The new queen, in her cage, is inserted in the center of the brood chamber between two combs. The cages in which queens are mailed by queen breeders are quite convenient for introducing queens. Those cages contain candy, and the queen is released by removing the cork or paper cover to enable the bees to eat through the candy, within the course of a day or two, to accomplish this.

It is well, when introducing a queen which has been received from a breeder, to allow the worker bees which accompany her to escape before placing the cage in the hive. The reason for this is that the newcomers are rarely accepted by the bees of the colony even after a day or two of confinement, except in the height of the honey harvest, while the queen is generally welcome after they have become acquainted with her.

During the height of the honey season a queen may be introduced more speedily by smoking the colony thoroughly after the removal of the old queen, closing it a few minutes, and allowing the new queen to run in at the entrance. Queens are most easily introduced when they are freshly removed from their hive and are in egg laying condition; so the transfer from one colony to another of a queen is attended with much less danger for her, than her introduction after she has arrived in a queen shipping cage.

After releasing the queen it is advisable to close the hive immediately, leaving it alone for a few days, for the greatest danger to the new queen comes from her being detected when robbers are flying about. Valuable queens should be introduced in a queen cage without attendants.

A surer way is by a "push-in" cage made of screen wire in which the queen is imprisoned. The "push-in" cage containing the queen is pushed into an area of emerging brood. Surrounded by the newly emerged bees, she is released safely by the bees eating out the candy of the cage. By this time she has acquired the odor of the hive and is readily accepted.

There is an absolutely safe method of introduction of a valuable queen. A hive or nucleus is filled with combs of emerging bees and the queen is released from the cage upon those combs. As there are no bees, except such as are in the process of emerging, there is no danger whatever for the queen. However, this should be attempted only in warm weather. Otherwise both unprotected brood and queen might become chilled and damaged.

The queen's wing may be clipped to prevent her from leaving with a swarm. In attempting to fly she will fall to the ground in front of the hive, and the bees, missing her, will return to the hive. Clipping must not be done until after the queen has met the drone or she will remain infertile. With a pair of sharp pointed scissors one of the front wings is lifted gently and about one-half of it cut off. She may be held by the wings or the thorax without danger of injuring her. She must not be held by the abdomen.

Some practical beekeepers always clip the wings of their queens. The queens are thus identified and their escape with the swarm is prevented. Also the swarm does not abscond, at least until a young queen is reared. But the apiary must be watched. Otherwise, the clipped queen may be lost; for she will attempt to fly and might be unable to return. In such cases she may usually be found accompanied by a small cluster of bees in the grass in front of the hive.

Dr. Miller advised, when clipping a queen, removing both of the wings on one side of the body. This makes the queen more easily visible among her bees, on account of her lopsided appearance.

Queens supplied by queen breeders are always expected to be properly mated and laying. Such queens are called untested queens. They are laying properly but not sufficient time has elapsed for bees to have emerged from the eggs; therefore, it is not positive that the queen is purely mated and that her offspring are even in color markings and characteristics.

Instead of buying queens, beekeepers often requeen their colonies from their best stock by inserting queen cells. The selected colony is dequeened and allowed to rear queen cells. When these queen cells are sealed they are cut out of the parent colony and introduced into the other colonies which have previously been deprived of their queens. In due course the young queen emerges from the newly introduced cell, goes out and is mated, and returns to the hive to head the colony.

There are some objections to this method. If the introduced cell is not accepted, the dequeened colony may rear its own cells; and since there are likely to be several of these cells, there is the added inducement for this colony to swarm. Another objection is that there is an interim of several days

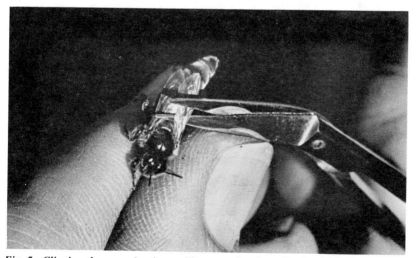

Fig. 5. Clipping the queen's wings will prevent her from flying and will also aid in identification. She may also be marked as in Fig. 2, p. 116.

between the introduction of the cell and the final return of the queen to take up her egg laying. During this time there are no eggs laid, and therefore no new brood is coming on. The progress of the colony is delayed just that much.

QUEEN-REARING

Modern queen-rearing, as practiced by the breeders who offer queens for sale, has come to be a profession of no little detail. Those who are interested will find books on the special subject of queen-rearing.

If the beekeeper has plenty of time and some fine stock either in his own apiary or that he has acquired from a queen breeder, he may attempt home queen-rearing.

As we have explained, the choice colony may be relieved of its queen. The colony immediately raises a number of queen cells. In eight or nine days these cells are sealed and ready to be removed to queen-rearing nuclei. Such queen-rearing can be done only at a season when bees will build and not tear down the cells and when it is sufficiently warm for the queens to fly around and become mated. This also makes necessary the rearing of plenty of drones in advance of the queen-rearing operations.

The nucleus is made by taking two or more frames, as may be desired (at least one of which should contain brood), with adhering bees and shaking into the hive the bees from one or more additional frames so that enough young bees remain, after the old bees have returned to their former hives, to their hives, to maintain the proper temperature for the sealed brood as well as to care for the emerging queen. In making up nuclei the queens from the old colonies must not be taken away with any of the frames.

It is advisable to use regular frames for nucleus hives, and to use either the ordinary hives with a division board or dummy to contract the brood chamber and economize the heat, or to make narrow hives just to suit the number of frames used.

These nuclei, having been made on the eighth or ninth day after the starting of queen cells, and in such number that there may be one queen cell for each and one for the mother colony, queen cells are cut out on the ninth or tenth day and one is inserted in each nucleus. The queen will soon emerge.

In cutting a queen cell, the comb is slit from each side of the base of the cell, not nearer than half an inch, upward to form a wedge-shaped piece with the cell in center. The base of the cell must not be squeezed or even handled. A similar wage-shaped piece is cut out of the frame of comb in which the cell is to be set, the cell being placed into the hole thus made, and fixed there securely. The frame is replaced in the hive and the hive closed.

After two or three days the queen will emerge, and a week or ten days later she will have been fertilized and will be laying. Eggs may be readily discovered in the cells.

As the virgin queen flies from the nucleus to meet a drone, sometimes the bees will accompany her if they have no unsealed brood. To prevent this, two or three days after the queens are emerged from the cells a frame containing eggs and young larvae is inserted in each nucleus. If, on her bridal tour, the queen should be lost, the unsealed brood will be on hand for the bees to use in rearing another to take her place.

Now the apiarist is ready for the formation of new colonies without the inconvenience of natural swarming.

Index

Don't Let Your Beekeeping Fall Behind the Times!

Subscribe now
to the...

AMERICAN BEE JOURNAL

The beekeeper's companion since 1861

SAVE $ BY READING THE AMERICAN BEE JOURNAL

Many subscribers say they easily save the cost of an entire year's subscription from a single informative article in the American Bee Journal. You can, too! Every issue of the Journal is packed with articles which appeal to a wide range of the beekeeping spectrum from the hobbyist to the professional.

INTERESTING AND UP-TO-DATE ARTICLES

Articles are selected from contributions sent to us from around the globe. Some are written by research entomologists, and commercial beekeepers, but many are written by the avid hobbyist just like you! Articles run the gamut from subjects important to us all like the Africanized honey bee, bee mites, practical management, honey marketing and promotion, all the way down to more specialized subjects such as how to make honey soft drinks, mead or how to create more enjoyment and profit from your beekeeping.

Published by Dadant & Sons,
Inc., Hamilton, IL 62341

1, 2 or 3 year subscriptions available.

Contact us for current prices.

Write for free sample copy and
booklist.